T0295194

IET ENERGY ENGINEERING SERIES 230

Self-Organizing Dynamic Agents for the Operation of Decentralized Smart Grids

Other volumes in this series:

Volume 1 **Power Circuit Breaker Theory and Design** C.H. Flurscheim (Editor)
Volume 4 **Industrial Microwave Heating** A.C. Metaxas and R.J. Meredith
Volume 7 **Insulators for High Voltages** J.S.T. Looms
Volume 8 **Variable Frequency AC Motor Drive Systems** D. Finney
Volume 10 **SF_6 Switchgear** H.M. Ryan and G.R. Jones
Volume 11 **Conduction and Induction Heating** E.J. Davies
Volume 13 **Statistical Techniques for High Voltage Engineering** W. Hauschild and W. Mosch
Volume 14 **Uninterruptible Power Supplies** J. Platts and J.D. St Aubyn (Editors)
Volume 15 **Digital Protection for Power Systems** A.T. Johns and S.K. Salman
Volume 16 **Electricity Economics and Planning** T.W. Berrie
Volume 18 **Vacuum Switchgear** A. Greenwood
Volume 19 **Electrical Safety: A guide to causes and prevention of hazards** J. Maxwell Adams
Volume 21 **Electricity Distribution Network Design, 2nd Edition** E. Lakervi and E.J. Holmes
Volume 22 **Artificial Intelligence Techniques in Power Systems** K. Warwick, A.O. Ekwue and R. Aggarwal (Editors)
Volume 24 **Power System Commissioning and Maintenance Practice** K. Harker
Volume 25 **Engineers' Handbook of Industrial Microwave Heating** R.J. Meredith
Volume 26 **Small Electric Motors** H. Moczala *et al.*
Volume 27 **AC–DC Power System Analysis** J. Arrillaga and B.C. Smith
Volume 29 **High Voltage Direct Current Transmission, 2nd Edition** J. Arrillaga
Volume 30 **Flexible AC Transmission Systems (FACTS)** Y-H. Song (Editor)
Volume 31 **Embedded Generation** N. Jenkins *et al.*
Volume 32 **High Voltage Engineering and Testing, 2nd Edition** H.M. Ryan (Editor)
Volume 33 **Overvoltage Protection of Low-Voltage Systems, Revised Edition** P. Hasse
Volume 36 **Voltage Quality in Electrical Power Systems** J. Schlabbach *et al.*
Volume 37 **Electrical Steels for Rotating Machines** P. Beckley
Volume 38 **The Electric Car: Development and future of battery, hybrid and fuel-cell cars** M. Westbrook
Volume 39 **Power Systems Electromagnetic Transients Simulation** J. Arrillaga and N. Watson
Volume 40 **Advances in High Voltage Engineering** M. Haddad and D. Warne
Volume 41 **Electrical Operation of Electrostatic Precipitators** K. Parker
Volume 43 **Thermal Power Plant Simulation and Control** D. Flynn
Volume 44 **Economic Evaluation of Projects in the Electricity Supply Industry** H. Khatib
Volume 45 **Propulsion Systems for Hybrid Vehicles** J. Miller
Volume 46 **Distribution Switchgear** S. Stewart
Volume 47 **Protection of Electricity Distribution Networks, 2nd Edition** J. Gers and E. Holmes
Volume 48 **Wood Pole Overhead Lines** B. Wareing
Volume 49 **Electric Fuses, 3rd Edition** A. Wright and G. Newbery
Volume 50 **Wind Power Integration: Connection and system operational aspects** B. Fox *et al.*
Volume 51 **Short Circuit Currents** J. Schlabbach
Volume 52 **Nuclear Power** J. Wood
Volume 53 **Condition Assessment of High Voltage Insulation in Power System Equipment** R.E. James and Q. Su
Volume 55 **Local Energy: Distributed generation of heat and power** J. Wood
Volume 56 **Condition Monitoring of Rotating Electrical Machines** P. Tavner, L. Ran, J. Penman and H. Sedding
Volume 57 **The Control Techniques Drives and Controls Handbook, 2nd Edition** B. Drury
Volume 58 **Lightning Protection** V. Cooray (Editor)
Volume 59 **Ultracapacitor Applications** J.M. Miller
Volume 62 **Lightning Electromagnetics** V. Cooray
Volume 63 **Energy Storage for Power Systems, 2nd Edition** A. Ter-Gazarian
Volume 65 **Protection of Electricity Distribution Networks, 3rd Edition** J. Gers
Volume 66 **High Voltage Engineering Testing, 3rd Edition** H. Ryan (Editor)
Volume 67 **Multicore Simulation of Power System Transients** F.M. Uriate
Volume 68 **Distribution System Analysis and Automation** J. Gers
Volume 69 **The Lightening Flash, 2nd Edition** V. Cooray (Editor)
Volume 70 **Economic Evaluation of Projects in the Electricity Supply Industry, 3rd Edition** H. Khatib
Volume 72 **Control Circuits in Power Electronics: Practical issues in design and implementation** M. Castilla (Editor)
Volume 73 **Wide Area Monitoring, Protection and Control Systems: The enabler for smarter grids** A. Vaccaro and A. Zobaa (Editors)
Volume 74 **Power Electronic Converters and Systems: Frontiers and applications** A.M. Trzynadlowski (Editor)
Volume 75 **Power Distribution Automation** B. Das (Editor)
Volume 76 **Power System Stability: Modelling, analysis and control** A.A. Sallam and B. Om P. Malik
Volume 78 **Numerical Analysis of Power System Transients and Dynamics** A. Ametani (Editor)
Volume 79 **Vehicle-to-Grid: Linking electric vehicles to the smart grid** J. Lu and J. Hossain (Editors)
Volume 81 **Cyber-Physical-Social Systems and Constructs in Electric Power Engineering** S. Suryanarayanan, R. Roche and T.M. Hansen (Editors)
Volume 82 **Periodic Control of Power Electronic Converters** F. Blaabjerg, K. Zhou, D. Wang and Y. Yang
Volume 86 **Advances in Power System Modelling, Control and Stability Analysis** F. Milano (Editor)
Volume 87 **Cogeneration: Technologies, optimisation and implementation** C.A. Frangopoulos (Editor)
Volume 88 **Smarter Energy: From smart metering to the smart grid** H. Sun, N. Hatziargyriou, H.V. Poor, L. Carpanini and M.A. Sánchez Fornié (Editors)
Volume 89 **Hydrogen Production, Separation and Purification for Energy** A. Basile, F. Dalena, J. Tong and T.N. Veziroğlu (Editors)

Volume 90 **Clean Energy Microgrids** S. Obara and J. Morel (Editors)
Volume 91 **Fuzzy Logic Control in Energy Systems with Design Applications in MATLAB®/Simulink®** İ.H. Altaş
Volume 92 **Power Quality in Future Electrical Power Systems** A.F. Zobaa and S.H.E.A. Aleem (Editors)
Volume 93 **Cogeneration and District Energy Systems: Modelling, analysis and optimization** M.A. Rosen and S. Koohi-Fayegh
Volume 94 **Introduction to the Smart Grid: Concepts, technologies and evolution** S.K. Salman
Volume 95 **Communication, Control and Security Challenges for the Smart Grid** S.M. Muyeen and S. Rahman (Editors)
Volume 96 **Industrial Power Systems with Distributed and Embedded Generation** R Belu
Volume 97 **Synchronized Phasor Measurements for Smart Grids** M.J.B. Reddy and D.K. Mohanta (Editors)
Volume 98 **Large Scale Grid Integration of Renewable Energy Sources** A. Moreno-Munoz (Editor)
Volume 100 **Modeling and Dynamic Behaviour of Hydropower Plants** N. Kishor and J. Fraile-Ardanuy (Editors)
Volume 101 **Methane and Hydrogen for Energy Storage** R. Carriveau and D.S-K. Ting
Volume 104 **Power Transformer Condition Monitoring and Diagnosis** A. Abu-Siada (Editor)
Volume 106 **Surface Passivation of Industrial Crystalline Silicon Solar Cells** J. John (Editor)
Volume 107 **Bifacial Photovoltaics: Technology, applications and economics** J. Libal and R. Kopecek (Editors)
Volume 108 **Fault Diagnosis of Induction Motors** J. Faiz, V. Ghorbanian and G. Joksimović
Volume 109 **Cooling of Rotating Electrical Machines: Fundamentals, modelling, testing and design** D. Staton, E. Chong, S. Pickering and A. Boglietti
Volume 110 **High Voltage Power Network Construction** K. Harker
Volume 111 **Energy Storage at Different Voltage Levels: Technology, integration, and market aspects** A.F. Zobaa, P.F. Ribeiro, S.H.A. Aleem and S.N. Afifi (Editors)
Volume 112 **Wireless Power Transfer: Theory, technology and application** N. Shinohara
Volume 114 **Lightning-Induced Effects in Electrical and Telecommunication Systems** Y. Baba and V.A. Rakov
Volume 115 **DC Distribution Systems and Microgrids** T. Dragicević, F. Blaabjerg and P. Wheeler
Volume 116 **Modelling and Simulation of HVDC Transmission** M. Han (Editor)
Volume 117 **Structural Control and Fault Detection of Wind Turbine Systems** H.R. Karimi
Volume 118 **Modelling and Simulation of Complex Power Systems** A. Monti and A. Benigni
Volume 119 **Thermal Power Plant Control and Instrumentation: The control of boilers and HRSGs, 2nd Edition** D. Lindsley, J. Grist and D. Parker
Volume 120 **Fault Diagnosis for Robust Inverter Power Drives** A. Ginart (Editor)
Volume 121 **Monitoring and Control Using Synchrophasors in Power Systems with Renewables** I. Kamwa and C. Lu (Editors)
Volume 123 **Power Systems Electromagnetic Transients Simulation, 2nd Edition** N. Watson and J. Arrillaga
Volume 124 **Power Market Transformation** B. Murray
Volume 125 **Wind Energy Modeling and Simulation, Volume 1: Atmosphere and plant** P. Veers (Editor)
Volume 126 **Diagnosis and Fault Tolerance of Electrical Machines, Power Electronics and Drives** A.J.M. Cardoso
Volume 127 **Lightning Electromagnetics, Volume 1: Return stroke modelling and electromagnetic radiation, 2nd Edition** V. Cooray, F. Rachidi, M. Rubinstein (Editors)
Volume 127 **Lightning Electromagnetics, Volume 2: Electrical processes and effects, 2nd Edition** V. Cooray, F. Rachidi, M. Rubinstein (Editors)
Volume 128 **Characterization of Wide Bandgap Power Semiconductor Devices** F. Wang, Z. Zhang and E.A. Jones
Volume 129 **Renewable Energy from the Oceans: From wave, tidal and gradient systems to offshore wind and solar** D. Coiro and T. Sant (Editors)
Volume 130 **Wind and Solar Based Energy Systems for Communities** R. Carriveau and D.S-K. Ting (Editors)
Volume 131 **Metaheuristic Optimization in Power Engineering** J. Radosavljević
Volume 132 **Power Line Communication Systems for Smart Grids** I.R.S. Casella and A. Anpalagan
Volume 134 **Hydrogen Passivation and Laser Doping for Silicon Solar Cells** B. Hallam and C. Chan (Editors)
Volume 139 **Variability, Scalability and Stability of Microgrids** S.M. Muyeen, S.M. Islam and F. Blaabjerg (Editors)
Volume 142A **Wind Turbine System Design, Volume 1: Nacelles, drive trains and verification** J. Wenske (Editor)
Volume 142B **Wind Turbine System Design, Volume 2: Electrical systems, grid integration, control and monitoring** J. Wenske (Editor)
Volume 143 **Medium Voltage DC System Architectures** B. Grainger and R.D. Doncker (Editors)
Volume 145 **Condition Monitoring of Rotating Electrical Machines** P. Tavner, L. Ran, C. Crabtree
Volume 146 **Energy Storage for Power Systems, 3rd Edition** A.G. Ter-Gazarian
Volume 147 **Distribution Systems Analysis and Automation, 2nd Edition** J. Gers
Volume 151 **SiC Power Module Design: Performance, robustness and reliability** A. Castellazzi and A. Irace (Editors)
Volume 152 **Power Electronic Devices: Applications, failure mechanisms and reliability** F Iannuzzo (Editor)
Volume 153 **Signal Processing for Fault Detection and Diagnosis in Electric Machines and Systems** M. Benbouzid (Editor)
Volume 155 **Energy Generation and Efficiency Technologies for Green Residential Buildings** D. Ting and R. Carriveau (Editors)
Volume 156 **Lithium-ion Batteries Enabled by Silicon Anodes** C. Ban and K. Xu (Editors)
Volume 157 **Electrical Steels, 2 Volumes** A. Moses, K. Jenkins, Philip Anderson and H. Stanbury
Volume 158 **Advanced Dielectric Materials for Electrostatic Capacitors** Q Li (Editor)
Volume 159 **Transforming the Grid Towards Fully Renewable Energy** O. Probst, S. Castellanos and R. Palacios (Editors)
Volume 160 **Microgrids for Rural Areas: Research and case studies** R.K. Chauhan, K. Chauhan and S.N. Singh (Editors)

Volume 161 **Artificial Intelligence for Smarter Power Systems: Fuzzy logic and neural networks** M.G. Simoes
Volume 165 **Digital Protection for Power Systems, 2nd Edition** Salman K Salman
Volume 166 **Advanced Characterization of Thin Film Solar Cells** N. Haegel and M Al-Jassim (Editors)
Volume 167 **Power Grids with Renewable Energy: Storage, integration and digitalization** A.A. Sallam and B. Om P. Malik
Volume 169 **Small Wind and Hydrokinetic Turbines** P. Clausen, J. Whale and D. Wood (Editors)
Volume 170 **Reliability of Power Electronics Converters for Solar Photovoltaic Applications** F. Blaabjerg, A. Haque, H. Wang, Z. Abdin Jaffery and Y. Yang (Editors)
Volume 171 **Utility-scale Wind Turbines and Wind Farms** A. Vasel-Be-Hagh and D.S-K. Ting
Volume 172 **Lightning Interaction with Power Systems, 2 Volumes** A. Piantini (Editor)
Volume 174 **Silicon Solar Cell Metallization and Module Technology** T. Dullweber (Editor)
Volume 175 **n-Type Crystalline Silicon Photovoltaics: Technology, applications and economics** D. Munoz and R. Kopecek (Editors)
Volume 178 **Integrated Motor Drives** X. Deng and B. Mecrow (Editors)
Volume 180 **Protection of Electricity Distribution Networks, 4th Edition** J. Gers and E. Holmes
Volume 182 **Surge Protection for Low Voltage Systems** A. Rousseau (Editor)
Volume 184 **Compressed Air Energy Storage: Types, systems and applications** D. Ting and J. Stagner
Volume 186 **Synchronous Reluctance Machines: Analysis, optimization and applications** N. Bianchi, C. Babetto and G. Bacco
Volume 190 **Synchrophasor Technology: Real-time operation of power networks** N. Kishor and S.R. Mohanty (Editors)
Volume 191 **Electric Fuses: Fundamentals and new applications, 4th Edition** N. Nurse, A. Wright and P.G. Newbery
Volume 193 **Overhead Electric Power Lines: Theory and practice** S. Chattopadhyay and A. Das
Volume 194 **Offshore Wind Power Reliability, availability and maintenance, 2nd Edition** P. Tavner
Volume 196 **Cyber Security for Microgrids** S. Sahoo, F. Blaajberg and T. Dragicevic
Volume 198 **Battery Management Systems and Inductive Balancing** A. Van den Bossche and A. Farzan Moghaddam
Volume 199 **Model Predictive Control for Microgrids: From power electronic converters to energy management** J. Hu, J.M. Guerrero and S. Islam
Volume 201 **Design, Control and Monitoring of Tidal Stream Turbine Systems** M. Benbouzid (Editor)
Volume 202 **Hydrogen from Seawater Splitting: Technology and outlook** A. Ray and M. Patel (Editors)
Volume 204 **Electromagnetic Transients in Large HV Cable Networks: Modeling and calculations** A. Ametani, H. Xue, T. Ohno and H. Khalilnezhad
Volume 208 **Nanogrids and Picogrids and Their Integration with Electric Vehicles** S. Chattopadhyay
Volume 210 **Superconducting Magnetic Energy Storage in Power Grids** M.H. Ali
Volume 211 **Blockchain Technology for Smart Grids: Implementation, management and security** H.L. Gururaj, K.V. Ravi, F. Flammini, H. Lin, B. Goutham, K.B.R. Sunil and C. Sivapragash
Volume 212 **Battery State Estimation: Methods and models** S. Wang
Volume 215 **Industrial Demand Response: Methods, best practices, case studies, and applications** H.H. Alhelou, A. Moreno-Muñoz and P. Siano (Editors)
Volume 213 **Wide Area Monitoring of Interconnected Power Systems, 2nd Edition** A.R. Messina
Volume 214 **Distributed Energy Storage in Urban Smart Grids** P.F. Ribeiro and R.S. Salles (Editors)
Volume 217 **Advances in Power System Modelling, Control and Stability Analysis, 2nd Edition** F. Milano (Editor)
Volume 225 **Fusion--Fission Hybrid Nuclear Reactors: For enhanced nuclear fuel utilization and radioactive waste reduction** W.M. Stacey
Volume 227 **Processing and Manufacturing of Electrodes for Lithium-Ion Batteries** J. Li and C. Jin (Editors)
Volume 228 **Digital Technologies for Solar Photovoltaic Systems: From general to rural and remote installations** S. Motahhir (Editor)
Volume 235 **Clean Energy Past and Future** P. Tavner
Volume 236 **Green HV Switching Technologies for Modern Power Networks** K. Niayesh (Editor)
Volume 237 **High Voltage Power Transformers: State of the art and technological innovations** M. Sforna, S. Zunino and F. Ferrari
Volume 238 **AI for Status Monitoring of Utility Scale Batteries** S. Wang, K. Liu, Y. Wang, D. Stroe, C. Fernandez and J.M. Guerrero
Volume 239 **Intelligent Control of Medium and High Power Converters** M. Bendaoud, Y. Maleh and S. Padmanaban (Editors)
Volume 241A **Power Electronic Converters and Systems, 2nd Edition, Volume 1: Converters and machine drives** M. Simões and T. Busarello (Editors)
Volume 241B **Power Electronic Converters and Systems, 2nd Edition, Volume 2: Applications** M. Simões and T. Busarello (Editors)
Volume 247 **Power System Strength: Evaluation methods, best practice, case studies, and applications** H. Alhelou, N. Hosseinzadeh and B. Bahrani (Editors)
Volume 250 **Interactions of Wind Turbines with Aviation Radio and Radar Systems** Alan Collinson
Volume 256 **Wireless Power Transfer Technologies: Theory and technologies, 2nd Edition** N. Shinohara (Editor)
Volume 905 **Power System Protection, 4 Volumes**

Self-Organizing Dynamic Agents for the Operation of Decentralized Smart Grids

Alfredo Vaccaro

The Institution of Engineering and Technology

Published by The Institution of Engineering and Technology, London, United Kingdom

The Institution of Engineering and Technology is registered as a Charity in England & Wales (no. 211014) and Scotland (no. SC038698).

First published 2024

The Institution of Engineering and Technology
Futures Place
Kings Way Stevenage
Hertfordshire, SG1 2UA, United Kingdom

www.theiet.org

British Library Cataloguing in Publication Data
A catalogue record for this product is available from the British Library

ISBN: 978-1-83953-687-8 (hardback)
ISBN: 978-1-83953-688-5 (PDF)

Typeset in India by MPS Limited

Cover image: Yaorusheng/Moment via Getty Images

Contents

About the author

Alfredo Vaccaro is a full professor at the Department of Engineering of the University of Sannio, Italy, where he is the chair of the Power System Research Group and dean of the Bachelor of Science and Master of Science in Energy Engineering.

He is the chair of the IEEE Power System Operation, Planning and Economics Committee—Technologies and Innovation Subcommittee (PSOPE-T&I), editor-in-chief of *Smart Grids and Sustainable Energy* published by Springer Nature, and associate editor of *IEEE Transactions on Power Systems* and *IEEE Transactions on Smart Grids*.

Chapter 1

Emerging needs and open problems in smart grids operation

Electrical power systems are considered to be among the most complex systems globally. However, these large-scale interconnected networks frequently operate under challenging conditions, which are mainly due to infrastructure improvement complexities, strict economic and environmental constraints, and various other technical factors, such as coordination among multiple system operators, heterogeneous grid standards, overlapping geographical regions, and monolithic operation tools.

In particular, the conventional control and monitoring paradigms currently adopted in power system operation tools should be radically enhanced in order to support the transition of existing electrical grids to fully decarbonized systems, where the large conventional power generators will be replaced by small-scale renewable power generators, which are characterized by random power injections distributed along the entire interconnected network. This transition process amplifies the complexities in managing both transmission and distribution systems, requiring larger spinning reserve margins, enhanced voltage regulation strategies, new flexibility sources, and improved coordination between transmission and distribution system operators (TSOs/DSOs).

Furthermore, the widespread development of deregulated energy markets has introduced additional complexities due to the dissolution of the traditional boundaries between the national power systems, which enables large power transactions throughout any buses of the interconnected power network, introducing new and correlated uncertainties source in short-term power system operation, and making power systems more susceptible to both internal and external disturbances.

The upgrade of transmission and distribution infrastructures in the task of increasing the grid meshing, raising the hosting capacity of distributed and dispersed generation systems, is another complex issue to address in the context of modern power systems, due to the complex and time-consuming procedures required for planning new infrastructures, which should adhere to stricter social and environmental constraints. These difficulties are pushing system operators in developing new tools aimed at maximizing the utilization of existing assets, differing the investments required for constructing new grid facilities.

Additionally, the operation of modern electrical grids will become more rigorous due to evolving system functions and rising demands for power quality and reliability.

As a consequence of these effects, several instances of wide-area blackouts, system separation, and large dynamic perturbation have been observed in many geographic area. These events are expected to become more frequent due to decreased security margins, reduced power system inertia, raising complexities in predicting market operators behavior, and the increasing randomness induced by a massive development of non-programmable energy sources.

Addressing these complexities by the smart grids technologies emerges as a promising solution. Indeed these technologies allow integrating computing, networking, and physical processes in the task of coordinating the grid resources holistically, with the final goal of enhancing power system efficiency, reliability, and the use of clean energy resources.

Central to smart grids technologies is the capability of distributed entities to collect, process, and share heterogeneous data. Thus, deploying flexible, reliable, and scalable computing architectures aimed at converting large data-streams into high-level information for decision-making is pivotal for the deployment of these technologies. In this context, the progress of information and communication technologies (ICTs) offers viable control and monitoring paradigms, which could enable pervasive information processing, supporting the development of advanced smart grid functions.

With this perspective in mind, many TSOs/DSOs worldwide have developed comprehensive experimental initiatives focused on implementing advanced ICT-based smart grid functions. These encompass synchronized measurement systems, adaptive protective strategies, on-line voltage regulation, demand-side management, knowledge extraction from massive data-sets, and wide-area monitoring and control systems [1,2].

While these experiences have demonstrated the effectiveness of ICT in enhancing the performance of operation tools for power transmission and distribution systems, they have been tailored to specific functions, such as data collection or information handling. They do not, however, encompass the concept of pervasive computing, which holds significant potential for managing the growing uncertainties inherent in modern transmission and distribution networks by enabling the solution of complex control and monitoring problems within defined time limits [3].

To tackle these challenges, the deployment of decentralized and cooperative computing architectures has been recognized as an enabling technology for adaptive control, distributed estimation, and pervasive monitoring of transmission and distribution grids [4]. The underlying concept is to devise a self-organizing framework that relies on a network of interactive smart sensors/controllers, which empower the distributed computing resources to solve complex operation problems by local cooperation, hence gaining problem-solving capabilities superior to the mere sum of the individual sensor/controller computing capabilities. The distributed sensors/controllers are based on modular, self-contained applications that can be remotely invoked via standard communication protocols, hence, ensuring the independence from both hardware/software platforms and the programming language employed for implementing the specific power system operation task [5]. This characteristic facilitates the enhancement of interoperability and integration levels for monolithic and difficult-to-customize power system control and monitoring functions. Furthermore, it enables computing tools to dynamically adapt to evolving changes in the structure and implementation of each grid operation function. It also allows for content-based data extraction and aggregation from sensors network, as well as the utilization of distributed embedded computing resources for decision-making support.

1.1 R&D challenges

The Energy Roadmap to 2050 defined by the European Commission represents a significant stride toward establishing a competitive low-carbon economy. The focal point of this strategic initiative revolves around the priority target of decarbonizing the energy systems. Achieving this ambitious goal hinges largely on the adoption and integration of generation systems powered by renewable energy sources. Notably, the large-scale deployment of renewable power generators, which are mainly dispersed at the power distribution level, will change the power system production paradigm from centralized to distributed and dispersed generation.

In this context, the pivotal role of distribution network management emerges as a linchpin in the overarching pursuit of global decarbonization objectives. Distribution networks, which will be one of the most important components of the new decarbonized power systems, assume the role of "active loads" for the transmission system. Hence, the critical issues characterizing power distribution systems operation will be intricately interwoven with the behavior and dynamics of transmission networks.

From this perspective, the deployment of smart distribution systems becomes imperative to facilitate the effective transition of existing power grids toward decarbonized systems. Indeed, distribution grids play a crucial role in real-time monitoring and control of the power flows from bulk power generation, through transmission systems, to the end-users. For this purpose, the development of smart operation tools for information and communication processing in power distribution grids is considered as the most promising enabling technology for enhancing the hosting capacity of low-carbon technologies, increasing the grid resilience, and raising the system reliability and security.

Unlike many other low-carbon energy technologies, the integration of smart tools in power distribution management systems presents a unique challenge. It necessitates implementation not only in existing distribution grids, some of which have surpassed the four-decade mark in many countries, but also in entirely new power systems. This process poses significant research and development challenges and requires overcoming various technical, economical, and social barriers. To facilitate this transition, governments play a crucial role in establishing clear, consistent policies, regulations, and plans for electricity systems. These measures are essential to encourage innovative investments in distribution networks. Additionally, fostering public engagement is paramount, involving education for all stakeholders, with a special emphasis on customers and prosumers. This outreach aims to convey the importance of smart distribution networks and the potential benefits they offer. Moreover, the effective deployment of smart distribution grids requires collaborative efforts among governments, research organizations, industry players, the financial sector, and international organizations.

From a technical perspective, numerous challenges need to be addressed. In particular, coordinating the heterogeneous and competitive objectives of generation companies, grid operators, end-users, and stakeholders in the electricity market is essential for reliable power distribution system operation. This coordination should aim to minimize costs and environmental impacts while maximizing system reliability, resilience, and stability. The overarching context is marked by aging infrastructures, persistent demand growth, the integration of an increasing number of variable renewable energy sources, the rise of electric vehicles, the need for enhancing the power system security, and reducing the overall carbon emissions.

Addressing these challenges requires a consistent focus on research and development efforts. Ongoing research should provide effective tools and technological options aimed at overcoming the mentioned hurdles. Specifically, efforts should concentrate on finding reliable and cost-effective solutions that significantly impact the effective development and deployment of smart distribution tools. These technology areas encompass generation, transmission, distribution, and the various range of electricity users, as discussed in the following subsections.

1.2 Renewable power generators

The main characteristic of renewable power generators lies in their inherent randomness, which is strongly related to the real-time variability of their primary energy sources, typically solar and wind. Indeed, the availability of these sources exhibits significant fluctuations throughout the day and the year, which are primarily driven by ever-changing weather conditions.

Another complex feature characterizing these generation systems is the uncontrollability of the generated power, which stems from its reliance on unpredictable weather patterns. Indeed, power output and operational times of solar and wind generators cannot be easily controlled by the generation companies. While there is the technical possibility to reduce the power output in particular operation states, this approach could be not economical feasible due to the large wastage of the primary source, and the sensible reduction of the generation conversion efficiency.

In contrast, traditional bulk thermal and hydropower plants are termed "dispatchable" due to their finely controllable operating times and power outputs. This important feature allows for precise power scheduling and real-time adjustments based on the power system operation state.

Linked closely to uncontrollability is the uncertainty affecting solar and wind power generators, which are mainly induced by the environmental phenomena affecting the energy conversion process, such as the short-term variability of clouds, and the sudden changes in wind speed and directions. These complex phenomena affect the intermittency and variability of the power generated by solar and wind generators, posing challenges to modeling and forecasting of the corresponding generated profiles.

Another important feature characterizing renewable power generators is their distributed nature. Unlike large thermal power plants, which are typically installed in remote locations and connected to transmission networks, renewable generators are mainly based on low-power generation units, which are spread across the entire power system, from high-voltage transmission to low-voltage distribution networks.

Most notably, the large-scale deployment of renewable generators in distribution networks induces critical technical issues, such as the bidirectional power flows, which sensibly raise the complexity of controlling, protecting and operating these grids, which have been planned assuming that all the electrical buses absorb power from the distribution grid (i.e. known as the "passivity hypothesis") and that the power flow direction always goes from the transmission to distribution networks (i.e. unidirectional power flow hypothesis). This paradigm shift requires revising the conventional criteria, methods and tools deployed in power distribution system planning, design and operation, increasing the responsibilities of distribution system operators (DSO), who should be able to manage active power systems, coordinating the operation of the distribution grids with the transmission system operators (TSO).

Finally, renewable power generators are frequently dispersed along large geographical area, since their location is dictated by the availability of the primary energy sources, which may not align with power system area experiencing the highest power demand. This spatial constraint poses complex technical challenges, necessitating the upgrade of the conventional control strategies (e.g. especially in voltage regulation) and grid improvements aimed at increasing the network meshing. Failure to implement these enhancements may affect the power quality and the system stability, resulting in adverse side effects such as curtailed renewable power generation, lower grid reliability, and higher power losses.

1.3 Effects of renewable power generators on distribution grids

The large-scale deployment of renewable power generators into power distribution systems necessitates addressing numerous technical issues, considering the features that have been analyzed before. A noteworthy categorization of these challenging issues arises from the fundamental performance requirements of power systems that, as outlined in [6], can be classified into power quality, power balance, power flow, and power system stability.

- Power quality: it pertains to the ideal nature of voltage and current signals. In an ideal scenario, electrical measurements within the network should yield perfect sinusoidal waves at a frequency of 50 Hz. However, this ideal condition does not hold true for the actual system. The growing prevalence of power electronics and particular power system operation states may contribute to a decline in power quality. Renewable power generators, which do not integrate any rotating machine, necessitate the use of power converter-based grid interface. This interface introduces frequency harmonics into the system due to the high switching frequency involved in power conversion. These harmonics result in power losses, inaccurate measurements, and additional technical challenges, particularly on the consumer side.

- Power balance: the concept of short- and long-term power balance pertains to maintaining a balance between the generated and demanded electric power. In the long term (i.e. multi-year time scenario), the power system ability to meet the forecasted power demand, which evolves over time, becomes crucial. For instance, with the expected surge in electric vehicle deployment in the near future, power systems need to properly adapt by promoting the installation of new energy capacity to handle the projected increase in load. It is imperative that the generation capacity consistently exceeds the maximum supply, incorporating a proper reserve margin aimed at managing the effects of system contingencies (e.g. multiple generation units go offline due to components faults). Renewable power generators are given dispatch priority on the electricity market due to their low marginal price, resulting in diminished revenues for thermal power plants and their subsequent uncommitment, thereby reducing the available energy capacity of the power system. This compelled decommissioning, stemming from reduced revenues, poses challenges to the long-term balance of power systems and necessitates the solution of both technical and economic issues. In contrast, short-term balance refers to satisfying real-time demand, ensuring that the daily load is continuously met. In this context, inaccurate forecasting of the power profiles generated by renewable generators can sensibly raise the complexities of real-time power balancing.
- Power flow: the decentralized characteristics of renewable power generators, along with their limited control of the generated power profile, have given rise to challenges in power flow management. These challenges include bidirectional flows and grid contingencies management. Power system contingencies are a natural outcome of the mismatch between the locations of power generation and consumption. For instance, in some areas of the Italian power system, a large quantities of electrical power generated by wind and solar generators, which are mainly located in the southern regions, cannot be fully exported due to congestion of some critical transmission lines. Contingencies result in both economic and environmental consequences, underscoring the importance of their prevention. On the other hand, bidirectional power flows can pose challenging issues to power system operation, which are mainly related to protection schemes, voltage regulation, and, in extreme cases, to grid islanding triggering cascading failures.
- Power system stability: it refers to the transient response of the power system following a disturbance (e.g. contingency event). Power systems typically operate in a quasi-steady state, transitioning from one equilibrium point to another through incremental adjustments. When a contingency occurs in the power grid, it introduces a transient behavior in the power system before it stabilizes at a new equilibrium point. Monitoring this transient behavior is crucial to prevent the system from reaching an unstable equilibrium. The transient state

is characterized by oscillations of varying amplitudes, depending on the power system operating point and physical parameters. Stability is achieved when these oscillations diminish over time, whereas instability arises when the oscillations exhibit a continual increase in amplitude, potentially leading to the shutdown of parts of the power system or even a blackout. Current stability measures involve controls like automatic voltage regulators or power system stabilizers, typically integrated with synchronous machines. However, with the increasing prominence of renewable power generators displacing thermal power plants, the reduction in the number of these controls raises concerns about system instability.

1.3.1 Power quality

Conventional power systems were primarily focused on delivering electric energy to end users at an acceptable voltage level. However, with the broader utilization of electric energy and the increasing deployment of modern electrical loads, the system requirements became more complex.

In an ideal scenario, customers would perceive the supply side as an ideal voltage source with zero impedance, characterized by a perfect 50 Hz waveform with a constant amplitude at any current level. However, achieving this ideal is not feasible in reality, and the field of power quality aims to investigate the difference between actual performance and ideal behavior [7].

The difference from the ideal operating conditions occurs under various circumstances. Disturbances related to power quality can be broadly categorized into two main types: variations and events. Variations involve slight deviations in voltage and current signals from their ideal values, which may be attributed to factors such as slow changes in loading conditions, harmonic distortion from converter-interfaced generation, and load. While these deviations have a small amplitude and a subtle departure from the ideal case, they are ever-present in power systems and should be managed up to a certain threshold. In particular, harmonics refer to the decomposition of power system signals into a combination of sinusoidal components with varying frequencies. Specifically, any periodic signal can be decomposed into a fundamental sinusoidal component at nominal grid frequency and a series of harmonic components, which are sinusoidal signals at frequency multiples of the nominal grid frequency.

Harmonic distortion can manifest as interharmonic distortions when the signal is non-periodic. Interharmonic distortions consist of waveforms whose frequencies are not multiples of the fundamental component. Subharmonic components are those with frequencies lower than the fundamental component, and they are treated separately due to the distinct effects they have on the grid.

Voltage flickers involve fluctuations in voltage between 1 and 15 Hz, potentially causing variations in light intensity and discomfort. Noise encompasses all other non-periodic frequency components.

The primary sources of harmonics are non-linear loads and power electronics, which have become more prevalent with the increasing integration of renewable power generators into power systems. These harmonics are introduced through the frequency switching of the power converter–based grid interfaces of these generating units.

The consequences of harmonics are significant for both voltage and current distortion. Harmonic voltage distortion can result in additional losses or voltage imbalances that may damage electrical equipment. Lower-order harmonics predominantly impact rotating machines, while higher-order ones affect capacitor banks. Electronic loads are notably susceptible to these distortions.

Conversely, harmonic current distortion can lead to the overheating of series components such as transformers and cables. The degree of heating increases with the distortion level. When the current exhibits high distortion, it may necessitate the de-rating of transformers, and a similar, albeit less pronounced, effect can be observed in cables and lines.

Other important phenomena inducing power quality issues are related to events encompassing larger and occasional deviations, such as those arising from alterations in network topology or grid faults. Voltage dips are examples of such events, characterized by their instantaneous nature rather than continuous occurrence, making them more challenging to predict and mitigate. Voltage dips represent issues in power quality arising from a brief reduction in voltage magnitude, leading to elevated currents at various locations. Motor starting, transformers, and faults are potential triggers for such events. Even if the reduction of the voltage magnitude due to voltage dips does not directly impact equipment or protective systems, it can trigger transient stability problems by diminishing the power transfer capability. Moreover, inverters and rectifiers within the grid may be adversely affected by voltage dips.

The frequency of voltage dips within a network is contingent on its structure and operation, ranging from a few occurrences to several hundred instances annually. To characterize a voltage dip, several aspects of the signal must be taken into account, including the start and end times of the dip, the minimum voltage magnitude recorded during the event, the depth influenced by the voltage before the dip, the duration of the dip, and the phase angle shift.

Renewable power generators play a significant role in both power system variations and events. For instance, in terms of frequency harmonics, the high switching frequency of converters introduces unwanted harmonics into the grid, disrupting the normal operation of power systems. Additionally, the intermittency of certain sources, like the power produced by photovoltaic power plants in the presence of clouds, leads to substantial variations in bus voltage magnitudes, causing perturbations on the entire power grid. A comprehensive review of these impacts is presented in [8].

In particular, wind turbines pose challenges to power quality within power systems, necessitating the accurate assessment of the harmonics of the voltage and current at the point of grid connection. In this context, the IEC Standard 61000-4-7 [9] serves as a benchmark for assessing the impacts of harmonics in wind turbines, advocating the use of Fourier transform with a rectangular window spanning 10 fundamental frequency periods, providing a 5 Hz resolution. According to [10], modern wind turbines exhibit a frequency spectrum extending up to a few kHz, with large generators displaying harmonic distortion even at lower frequencies.

Photovoltaic generation systems could induce many power quality issues in power systems, as detailed in [11]. In particular, cloud shading in these systems can result in very-short voltage variations, which can significantly impact the bus voltage magnitudes, especially in rural networks. These voltage fluctuations may induce flicker, leading to additional stress on electrical equipment, and potentially triggering malfunctions in critical network components. Another source of voltage variations in photovoltaic systems is associated with faulty or badly designed maximum power point tracking system, which aims at enhancing the energy conversion efficiency. Further challenges arise from current transients during start-up and shut-down in low irradiation operation states, influenced by the inverter dynamics and its transient characteristics. These conditions also contribute to high total harmonic distortion during periods of low solar radiation.

1.3.2 Power balance

An additional significant challenge posed by the operation of a power system with a large penetration of renewable power generators pertains to power system balance, which require addressing complex challenging issues that can be categorized based on the time scale to which they pertain as

- long-term balance, which is associated with the adequacy problem;
- short-term balance, often denoted as regulation problems.

Solving these issues involves considerations in both electricity market design and technological aspects.

In particular, in the medium to long term, adequacy involves the power system's capability to meet demand in all potential configurations with an adequate margin. These configurations hinge on demand variability, the availability of generation capacity, transmission constraints, and imports from foreign power systems. Uncertainty regarding these configurations escalates with a broader observation period. For instance, forecasting the load profiles resulting from the electrification of transport is more manageable one year in advance compared to a decade.

While the challenge of adequacy has been a longstanding concern in power systems, it is growing more complex with the increasing decommissioning of thermal power plants, which is caused not only by the strictest environmental policies but also by the reduced economic profits deriving from trading energy in day-ahead markets, where the increasing number of energy transactions ruled by renewable power generators is settling the energy markets at lower equilibrium prices reducing the revenues for all market operators. The consequence of this phenomena is the decommissioning of less efficient thermal power plants, which reduces the power system adequacy and the grid flexibility.

Beyond the large-scale deployment of distributed power generators, the electrification of the transportation systems could further impact power system adequacy. In this context, the adequacy margin, representing the power available after meeting the power demand, is expected to decrease over time and, without implementing proper corrective measures, power systems are expected to operate in increasingly critical conditions. In particular, recent studies tried to quantify the impacts on power system adequacy deriving by the replacement of thermal power plants with renewable power generators, concluding that 100 MW of wind or solar power capacity is equivalent to about 11 MW of thermal capacity.

To alleviate the decline in the power system adequacy margin, the introduction of a quantitative index to measure this characteristic is adopted. In particular, in many European power grids, the system adequacy is assessed through the loss of load expectation (LOLE), with a power system considered adequate if the LOLE is less than 3 h per year.

To maintain this level of LOLE, interventions in the electricity market have been proposed to replace the decommissioned power capacity, e.g. coming from coal-fired power plants, with new low-carbon technologies. One potential solution involves establishing a capacity market, allowing thermal power plants to be remunerated independently of the electricity market, where their participation is expected to diminish due to the integration of renewable power generators and the resulting reduction in market equilibrium prices. As discussed in [12], energy-only markets, which rely solely on remuneration from electricity markets, cannot guarantee a long-term balance between supply and demand without accompanying measures as far as energy storage, capacity markets, and scarcity pricing are concerned.

As far as power balancing on short timescales is concerned, it involves aligning supply and demand from hours to seconds before real-time operations [13]. In this context, maintaining power balance is essential to prevent load shedding and ensure the power system stability. Intra-day markets and ancillary services play crucial roles in guaranteeing power balancing on this timescale.

The stochastic nature of renewable energy sources, particularly their dependence on weather conditions, affects short-term power balance. Coping with the growing uncertainty can be approached in two main ways within the market framework. One approach involves increasing reserves from balancing markets to manage the random power imbalances. The other approach entails imposing higher penalties on generators introducing imbalances due to inaccurate forecasting. The first solution incurs additional costs for the system operator, potentially elevating the cost of the balancing reserves, which increase the overall transmission costs. The second solution approach may discourage new investments in renewable power generators due to the risk of penalties resulting from inaccurate forecasting.

To distribute responsibilities effectively, a balanced combination of both approaches is recommended, since the reserve margins needed to reliably manage the power unbalances are assessed by the transmission system operators during the pre-dispatch phase, while the financial penalties are settled after imbalances occur.

The prevalent approach employed by TSOs to reliably assess the reserve margins is often deterministic. Reserves are sized under the assumption of a specific network event, such as the largest credible contingency, according to the N-1 reliability criteria. However, this approach tends to be overly conservative, leading to unnecessary costs. Alternative methods include probabilistic approaches, which necessitate detailed knowledge about potential sources of imbalance and their probability distribution. Moreover, deterministic reserve sizing is frequently established for an extended period, such as one year, known as static sizing. An alternative dynamic sizing approach involves periodic calculations of the required reserves based on the network's prevailing conditions.

To enhance short-term power system balance in the presence of renewable power generators, reducing forecasting errors by shifting the electricity markets closer to real-time operations is a viable solution. For instance, instead of balancing the system 1 h ahead, sub-hourly transactions can enable renewable power generators to accurately bidding the actual power generated. Another interesting option involves allowing distributed generators to participate in the ancillary markets, hence, offering grid services aimed at enhancing power system flexibility.

1.3.3 Power flow

The flow of electric energy within power systems is ruled by the theory of electrical circuits and depends on the power components parameters, the network topology, and the spatial distribution of generators and loads. Unfortunately, network operators face limitations in precisely controlling the directions and magnitudes of the power flows. Indeed, while power flows can be regulated to some extent through tools like phase-shifting transformers and flexible alternating current transmission systems, their control over the entire power grid is not technically feasible.

Congestion arises when the constraints on the transmission system are unable to accommodate all power flows within it [14]. The increasing integration of renewable power generators in power grids is contributing to a rise in network congestion, necessitating the implementation of advanced decision-support tools for congestion management.

Various factors limit the transmission capability of overhead lines. The initial analysis of the load capability of overhead lines traces back to 1953 with the work of Clair, who established the theoretical framework and categorized the main causes of capability limits: conductor thermal limits, voltage drop, and stability limits [15].

The thermal limit is linked to the type of conductor used and is the primary restricting factor for short lines, up to 80 km. This limit is related to the thermal elongation resulting from rising cable temperatures caused by solar irradiation and the Joule effect. Extended cables may come into contact with surrounding elements, leading to hazardous phase-to-ground faults. Once the maximum allowable temperature is determined, the maximum current circulating through the transmission line (i.e. the line ampacity) is calculated, considering the thermal balance between the cable and the environment.

The maximum allowable voltage drop is another relevant element that may constrain the maximum load capability of overhead lines, especially for line lengths ranging from 80 to 320 km. This phenomena is highly dependent on the power factor of the transmitted power, which is commonly assumed to be unitary in loading calculations. This assumption might not always hold true and should be reconsidered when comparing AC transmission lines to high-voltage direct current to avoid unfairly penalizing the latter solution.

Finally, as far as stability limits are concerned, they are often defined by considering steady-state stability margins, which primarily impacts long lines (over 320 km). Factoring in transient stability would extend the limit to shorter lines as well.

1.3.4 Power system stability

Power system stability is defined as "the ability of the system, for a given initial operating condition, to return to a state of operating equilibrium after a physical disturbance, with most of the variables of the system constrained so that the entire system remains intact" [16]. This definition underscores the system capacity to endure both minor disturbances, such as load variations, and, more significantly, major disturbances like generator failures or line short circuits.

To analyze dynamic events that could jeopardize power system stability and design effective strategies aimed at mitigating their effects, a comprehensive classification of power system stability is crucial. This classification should consider various features, including disturbance magnitude, time dynamics, involved devices, and, notably, the physical nature of the instability [16].

Recent revisions and expansions of stability classification have incorporated considerations for the grid effects of renewable power generators connected to the grid by power converter-based interfaces. In this context, two additional categories, namely resonance stability and converter-driven stability, have been introduced alongside the existing categories of rotor angle stability, frequency stability, and voltage stability. This expansion was necessary because, unlike conventional generators characterized by relatively slow electromechanical phenomena, converter-based grid interfaces exhibit much faster dynamics, ranging from a few microseconds to several milliseconds, giving rise to complex electromagnetic phenomena [17].

1.4 Enabling technologies for enhancing the hosting capacity of renewable power generators

Advancements in research and technology in the context of renewable power generators could significantly influence the potential for their large-scale integration, particularly within power distribution systems. The challenging issues to address are related to the intermittence of the generated power profiles, which may sensibly affect the correct operation of the electricity system. The most promising enabling technologies aimed at facing this challenging issue include [e-highway]:

- Energy storage systems: they offer a solution to alleviate the effects of the generation intermittency by partially decoupling the generation and delivery of the generated power. Automation of control of power generation and power demand, alongside with the deployment of demand-side management systems, can play a crucial role in achieving a balance between power supply and demand. In this context, there is a need for developing research activities in the domains of power system planning, generation forecasting, and the development of smart tools for real-time coordination of distributed energy resources.
- Wide-area monitoring protection and control systems: this technology has primarily been designed for interconnected power transmission systems, providing real-time monitoring and visualization of power system components and performance. It serves the purpose of assisting system operators in comprehending and optimizing the behavior and performance of power system elements across

interconnections and vast geographical areas. While the technology is also relevant to some extent at the distribution level, its application needs reinforcement at the points where transmission and distribution networks intersect. Research activities in this domain are concentrated on developing decision support tools aimed at enhancing security at the distribution level and facilitate the integration of renewable power generators. The deployment of wide-area monitoring, protection and control technologies, coupled with advanced analytic tools for knowledge discovery from big data-streams, could promptly generate valuable information that support decision-making, mitigates disturbances, and enhances distribution capacity and reliability. Researchers must strive to identify cost-effective solutions to make these capabilities viable at the distribution level.

- Information and communication technology: information and communication systems, whether utilizing dedicated utility communication networks or public carriers and networks, facilitates data transmission for both deferred and real-time operations, even during outages. In addition to researching communication devices, it is important to develop substantial investigation and development activities in the area of real-time computing, system control software, and enterprise resource planning software. This is necessary to facilitate the two-way exchange of information among stakeholders and enable a more efficient utilization and management of the distribution network.
- Distribution network automation and management: sensing and automation in power distribution systems have the potential to decrease outage and restoration duration, uphold voltage levels, and enhance asset management. In this context, the research initiatives have been focused on enhancing the distribution management systems, leveraging real-time data from monitoring relays for tasks such as fault location, automatic feeder reconfiguration, and optimization of voltage and reactive power. Additionally, these technologies can be deployed to control distributed and dispersed generation systems. The large pervasion of sensor technologies in distribution grids enables condition- and performance-based maintenance of network components, leading to optimized equipment performance and more effective asset utilization. In this context, research on both hardware and software technologies holds crucial importance.
- Distribution networks enhancement applications: many technologies and applications for distribution systems draw concepts directly from the field of transmission, where flexible AC transmission systems are deployed for the task of enhancing network controllability and maximize power transfer capability. The application of this technology on existing lines is known to enhance efficiency, deferring the necessity for new investments. Encouraging research activities focused on device development and modeling distribution networks is essential. Furthermore, DC technologies can find utility at the distribution level, particularly in power networks characterized by a massive penetration of renewable energy generation systems, which typically require power converter-based interfaces for grid interconnection. The implementation of DC technologies at this level aim at reducing power system losses and improve system reliability, enabling a more efficient utilization of energy sources.

- Electric vehicle charging infrastructure: power distribution systems will integrate a large number of electric vehicle charging infrastructures, which necessitate the management of billing, scheduling, and other smart features aimed at coordinating the charging/discharging profiles of electric vehicles fleets (i.e., according to the vehicle-to-grid paradigm). Ideally, charging operations should align with periods of low energy demand, and distribution system operators need to rigorously assess the potential grid impacts of both slow and fast charging stations in terms of cost/benefit ratios. In the long term, it is envisioned that large charging installations could offer power system ancillary services, such as capacity reserve, peak load shaving, and vehicle-to-grid regulation. Comprehensive cross-disciplinary analyses, covering technical and socio-economic aspects, are extremely relevant in this context.
- Advanced metering infrastructure: a smart distribution network, particularly within the context of electricity markets, should leverage an advanced metering infrastructure. This entails implementing various technologies that furnish customers and utility operators with information on electricity prices and consumption, encompassing details like the time and amount of electricity used. Engaging in research endeavors within this domain, involving the development of technology solutions and system models, is crucial for implementing smart grid functions. These functionalities include providing remote consumer price signals to offer time-of-use pricing information, the capacity to collect, store, and report customer energy consumption data in different time intervals or near real-time, enhanced energy diagnostics through more detailed load profiles, the ability to remotely identify the location and extent of outages by signaling when the meter goes out and when power is restored, remote connection and disconnection capabilities, and facilitating a retail energy service provider in managing revenues through more effective cash collection and management.

The large-scale deployment of these enabling technologies allows for addressing the following challenging issues:

1. Mitigate the adverse side effects derived from a massive pervasion of renewable power generators into power distribution networks, which could be obtained by:
   (a) Increasing the loading levels of power components.
      - Enhancing the renewable power and load forecasting tools.
      - Development of technologies aimed at increasing the load demand flexibility.
      - Development of demand-side management systems and demand response programs.
      - Increase the load capability of power components by developing advanced tools for dynamic loading (i.e., dynamic thermal rating systems)

   (b) Increasing the hosting capacity of renewable generators.
      - Enhancing the accuracy of generation capacity forecasting algorithms.
      - Increase the flexibility of the active/reactive power profiles generated by renewable power generators and coordinate their operation

in order to support the grid (e.g., primary frequency control and voltage regulation).
- Promotion of research activities on storage systems technologies.
- Large-scale deployment of energy storage systems and coordinate their operation in order to support the grid.

(c) Deployment of smart protective systems.
- Develop pervasive sensing systems for network diagnostic and condition monitoring.
- Enhancing distribution automation and self-healing (e.g. optimal network reconfiguration).
- Development of fast/high-performing electronic protection devices.
- Development of adaptive protection schemes.

(d) Development of smart tools for on-line voltage control.
- Development of generators power output control methods.
- Conceptualize mathematical frameworks aimed at coordinating a large number of dispersed energy resources in the task of providing ancillary grid services (e.g., active/reactive power flow control).
- Use of emerging technologies aimed at increasing the distribution grid flexibility (e.g., D-FACTS).
- Deployment of adaptive models for state estimation of power distribution system.

(e) Increasing the grid reliability and adequacy.
- Developing forecasting algorithms aimed at assessing the availability of renewable energy sources.
- Adaptive coordination of protection and control systems.
- Developing new tools for fault identification, isolation, and power system restoration.
- Developing smart tools aimed at coordinating the operation of distributed microgrids (e.g., multi-microgrids networks).
- Research on potential network topological changes (use of meshed networks, increase in substations number, distribution feeders length reduction, etc.)

2. Coordinating the operation of power distribution and transmission systems, which can be obtained by:

(a) Developing new tools for integrated management of transmission and distribution grids.
- Enhancing planning tools to a fully integrated network planning for both transmission and distribution grids.
- Developing new tools aimed at coordinating the operation of both transmission and distribution systems.
- Define standard interfaces aimed at allowing information sharing between transmission and distribution operation tools.
- Developing advanced algorithms aimed at forecasting the distribution networks behavior.

(b) Enhancing the interfacing inside electrical stations.
- Development of smart protective devices.
- Improvement of SCADA and automation systems.

- Development of new methods aimed at coordinating transmission and distribution automation tools.

3. Increase the hosting capacity of electric vehicles charging stations in distribution grids, which can be obtained by
   (a) Raising the security and reliability of power distribution grids.
   (b) Developing new tools aimed at coordinating the vehicle storage systems in the task of providing ancillary grid services.
   (c) Developing pervasive monitoring and distribution automation systems.
4. Enhancing the coordination between market and power grid operations, which can be obtained by
   (a) Increasing the penetration of multi-carrier energy networks.
   (b) Deployment of integrated planning methods.
   (c) Enhancing the interface capability of distribution grids with energy hubs.
   (d) Developing mathematical models aimed at assessing the infrastructures reliability and adequacy through energy conversion systems.
   (e) Developing adaptive algorithms for multiple energy forecasting on different time-scales.
   (f) Integrating energy hubs in demand response programs.
   (g) Development of smart tools for optimal energy flow management in multi-energy carrier networks.
5. Conceptualizing decentralized/not hierarchical control and monitoring architectures, which can be obtained by
   (a) Defining new technical guidelines for information processing in smart grids.
   (b) Increasing the redundancy of power system communication by deploying multi-channels communication units.
   (c) Developing effective and reliable routing algorithms for data exchange in power distribution grids.
   (d) Identify a suitable trade-off between centralized versus distributed control paradigms.
   (e) Upgrading the existing computing architectures in order to enhance their scalability and resilience to external perturbations.
   (f) Implements pervasive hardware and software countermeasures aimed at enhancing the cybersecurity of the computing resources.
6. Enable distributed state estimation of the power distribution grid, which can be obtained by
   (a) Defining standard interfaces among grid components, tools, and applications.
   (b) Enabling information sharing between heterogeneous tools.
   (c) Implementing middlewares for data sharing.
   (d) Define comprehensive guidelines for data synchronization (accuracy and availability).
   (e) Implements distributed mathematical tools for reliable state estimation.

7. Increase the pervasion of grid sensors, which can be obtained by
   (a) Deployment of high-throughput data and broadband services, which could be based on hybrid wired/wireless communication units.
   (b) Enhancing the scalability of existing sensing infrastructures.
   (c) Increasing the cybersecurity of grid sensors and monitoring tools.
8. Developing proactive protective systems, which can be obtained by
   (a) Defining protection schemes based on adaptive relays.
   (b) Developing reliable protection algorithms based on self-validated computing.
   (c) Deploying adaptive control algorithms.
   (d) Conceptualizing decentralized architectures for reliable information processing and data sharing between cooperative controllers.
   (e) Implementing adaptive forecasting models.
9. Developing advanced tools for big data storage and processing, which can be obtained by
   (a) Defining service-oriented architectures for smart grids data processing.
   (b) Deploying scalable computing frameworks for real-time computing.
   (c) Standardizing the semantics and developing taxonomies for ontology drives decision support tools.
   (d) Developing computing frameworks that are platform, language, and vendor independent.
   (e) Developing information processing tools for knowledge discovery and extraction from large heterogeneous sensor networks.
   (f) Conceptualization of middleware platforms able to effectively, efficiently, and easily aggregate different kinds of devices to
       - sense events from the smart grid;
       - analyze them by using computational intelligence-based techniques;
       - define control actions aimed at mitigating the potential impacts of the identified events.
10. Developing highly scalable computing frameworks for online power system analysis, which can be obtained by
    (a) Integrating a wide spectrum of communication technologies, which range from high-speed local area network-based systems, lower-speed wide area networks, wireless networks, and ad hoc wireless networks.
    (b) Deploying a quality of service-based brokering system aimed at satisfying fixed quality of service constraints (e.g., maximum computing time).
    (c) Deploying the power system models on processing platforms based on parallel computing or computational grids.
    (d) Developing static/dynamic modeling techniques based on validated simulation software.
    (e) Deploying reliable computing paradigms for uncertainty modeling and processing.
    (f) Implements standard software interfaces for data exchange between operation tools.

11. Develop cooperative and bio-inspired paradigms for self-healing power networks, which can be obtained by
    (a) Defining new technical guidelines for decentralized information processing in smart grids.
    (b) Increasing the pervasivity and the resilience of power system communication by deploying pervasive sensor networks, ubiquitous computing-based systems, and multi-channels communication units.
    (c) Deploying advanced computing paradigms aimed at handling complex systems by information semantics.
    (d) Identifying a suitable trade-off between centralized versus distributed control paradigms.
    (e) Upgrading of the existing computing infrastructures in order to enhance
        - Data sharing among applications.
        - Integration of functions among various tools.
        - Standardization of the data model.
        - Sharing of data and information among software functions.

1.5 Toward a self-organizing framework for smart grids computing

To address many of these R&D challenges, the conceptualization of decentralized and self-organizing computing architectures have been recognized as one of the most promising enabling technologies.

Developing an effective self-organizing framework for decentralized smart grids computing encounters several challenges, which mainly stem from the need to upgrade and establish interoperability for existing power system operation tools, which are often built on client–server platforms characterized by heterogeneous information technologies, data exchange interfaces, and information processing paradigms. Within these conventional tools, a large volume of raw data is typically collected by distributed grid sensors and then transmitted to a central server for post-processing analyses [18]. The implementation of this hierarchical paradigm within the smart grid context prompts a reassessment of numerous design principles and assumptions. These encompass scale, reliability, heterogeneity, manageability, and system evolution over time [19].

With dispersed generators being fundamental components of smart grids, challenges related to the scalability and reliability of the computing architectures come to the forefront. These encompass promptly analyzing large data-streams generated by pervasive grid sensors distributed at different voltage levels over the entire power network. Additionally, the heterogeneity of grid sensors, which was not an issue in conventional power distribution systems, now necessitates attention since as measurement systems expand, scaling with the same hardware and software technologies becomes challenging. Manageability also represents a relevant issue to address, given that smart grids could integrate hundreds or even thousands of grid sensors. Lastly, as smart grids evolve, the computing architecture should dynamically adapt.

Handling massive amounts of data emerges as another crucial challenge, as the number of grid measurements in smart grids is projected to grow exponentially [20], resulting in a sensible increase of the data-streams that should be promptly processed to infer actionable insights within very short time-frames [21]. In this context, the data processing functions need to represent and manage the inherent uncertainties affecting the measured data in order to correctly characterize the information context, assessing the degree of confidence of the processed information [22].

Furthermore, even when sophisticated mathematical models aimed at analyzing the data-streaming are available, various issues should be solved, such as communication network congestion, intractable mathematical problems, and the need for hardware redundancies in order to reduce the vulnerability of the centralized computing systems.

In light of these challenges, recent advancements in collaborative and cooperative computing have given rise to non-hierarchical, decentralized monitoring architectures. These architectures are based on reliable and highly scalable information processing paradigms. Notably, collaborative and decentralized architectures have been first proposed for voltage quality monitoring and regulation in smart grids [23,24]. Utilizing self-organizing sensor networks, these architectures enable local detection of voltage anomalies through information exchange among neighboring nodes. Additionally, various protocols have been examined to achieve non-hierarchical paradigms, such as an IEEE802.15.4-based wireless sensor network for urban-scale smart grid environments [4].

These research studies conceptualized a fully decentralized computing architecture based on a "think locally, act globally" approach. The central idea is based on the information-spreading theory for coordinating a network of self-organizing sensors/controllers responsible for executing control, monitoring, estimation, and detection functions [25,26]. In this context, each sensor/controller is characterized by the following features:

1. It is located in a specific power system bus.
2. It is directly connected with a limited number of local sensors/controllers by a short-range communication system.
3. It is equipped with a dynamic agent, which is modeled by a first-order dynamic system (oscillator), whose state is coupled with nearby oscillators through proper local coupling strategies.

By properly designing the coupling strategies between the dynamic agents, it is possible to allow the cooperative sensors/controllers to achieve decentralized consensus on general functions of all the local variables. This bio-inspired approach enables the sensors/controllers to solve complex operation problems by processing and sharing only local information. In particular, consensus among the sensors/controllers network allows time synchronization and estimation of global variables characterizing current power system operation, without relying on a centralized fusion center. As a result, the fundamental monitoring, control, and estimation of smart grids functions can be fully implemented by a decentralized/non-hierarchical framework, which offers advantages over traditional client–server paradigms, including higher scalability, reduced communication bandwidth, and enhanced extension and reconfiguration features.

References

[1] Cirio D, Lucarella D, Giannuzzi G, *et al.* Wide area monitoring in the Italian power system: architecture, functions and experiences. *European Transactions on Electrical Power.* 2011;21(4):1541–1556.

[2] Giannuzzi G, Lauria D, Pisani C, *et al.* Real-time tracking of electromechanical oscillations in ENTSO-e Continental European Synchronous Area. *International Journal of Electrical Power & Energy Systems.* 2015;64:1147–1158. Available from: https://www.sciencedirect.com/science/article/pii/S0142061514005675.

[3] Carlini EM, Giannuzzi GM, Mercogliano P, *et al.* A decentralized and proactive architecture based on the cyber physical system paradigm for smart transmission grids modelling, monitoring and control. *Technology and Economics of Smart Grids and Sustainable Energy.* 2016;1(5):1–5.

[4] de Mues MO, Alvarez A, Espinoza A, *et al.* Towards a distributed intelligent ICT architecture for the smart grid. In: *2011 9th IEEE International Conference on Industrial Informatics*; 2011. p. 745–749.

[5] Vaccaro A, Popov M, Villacci D, *et al.* An integrated framework for smart microgrids modeling, monitoring, control, communication, and verification. *Proceedings of the IEEE.* 2011;99(1):119–132.

[6] Sinsel SR, Riemke RL, and Hoffmann VH. Challenges and solution technologies for the integration of variable renewable energy sources – a review. *Renewable Energy.* 2020;145:2271–2285.

[7] Bollen MH. What is power quality? *Electric Power Systems Research.* 2003;66(1):5–14.

[8] Liang X. Emerging power quality challenges due to integration of renewable energy sources. *IEEE Transactions on Industry Applications.* 2016;53(2):855–866.

[9] Commission IIE, *et al.* IEC 61000-4-37. 2016.

[10] Tentzerakis ST and Papathanassiou SA. An investigation of the harmonic emissions of wind turbines. *IEEE Transactions on Energy Conversion.* 2007;22(1):150–158.

[11] Pakonen P, Hilden A, Suntio T, *et al.* Grid-connected PV power plant induced power quality problem: experimental evidence. In: *2016 18th European Conference on Power Electronics and Applications (EPE'16 ECCE Europe).* IEEE; 2016. p. 1–10.

[12] Yang J. Resource adequacy: economic and engineering challenges and proposed solutions. *IEEE Power and Energy Magazine.* 2006;4(2):59–65.

[13] Hirth L and Ziegenhagen I. Balancing power and variable renewables: three links. *Renewable and Sustainable Energy Reviews.* 2015;50:1035–1051.

[14] Jayaweera D. *Smart Power Systems and Renewable Energy System Integration.* New York, NY: Springer; 2016.

[15] Clair H. Practical concepts in capability and performance of transmission lines. *Transactions of the American Institute of Electrical Engineers. Part III: Power Apparatus and Systems*. 1953;72:1152–1157.
[16] Kundur P, Paserba J, Ajjarapu V, *et al.* Definition and classification of power system stability IEEE/CIGRE joint task force on stability terms and definitions. *IEEE Transactions on Power Systems*. 2004;19(3):1387–1401.
[17] Hatziargyriou N, Milanovic J, Rahmann C, *et al.* Definition and classification of power system stability—revisited and extended. *IEEE Transactions on Power Systems*. 2020;36(4):3271–3281.
[18] Giri J, Parashar M, Trehern J, *et al.* The situation room: control center analytics for enhanced situational awareness. *IEEE Power and Energy Magazine*. 2012;10(5):24–39.
[19] Chakrabarti S, Kyriakides E, Bi T, *et al.* Measurements get together. *IEEE Power and Energy Magazine*. 2009;7(1):41–49.
[20] Fan L, Li J, Pan Y, *et al.* Research and application of smart grid early warning decision platform based on big data analysis. In: *2019 4th International Conference on Intelligent Green Building and Smart Grid (IGBSG)*; 2019. p. 645–648.
[21] Zobaa AF, Vaccaro A, and Lai LL. Enabling technologies and methodologies for knowledge discovery and data mining in smart grids. *IEEE Transactions on Industrial Informatics*. 2016;12(2):820–823.
[22] IEEE Vision for Smart Grid Communications: 2030 and Beyond Reference Model. 2013;p. 1–11.
[23] Andreotti A, Petrillo A, Santini S, *et al.* A decentralized architecture based on cooperative dynamic agents for online voltage regulation in smart grids. *Energies*. 2019;12(7).
[24] Di Bisceglie M, Ullo SL, and Vaccaro A. The role of cooperative information spreading paradigms for smart grid monitoring. In: *2012 16th IEEE Mediterranean Electrotechnical Conference*; 2012. p. 814–817.
[25] Loia V and Vaccaro A. Decentralized economic dispatch in smart grids by self-organizing dynamic agents. *IEEE Transactions on Systems, Man, and Cybernetics: Systems*. 2014;44(4):397–408.
[26] Vaccaro A, Velotto G, and Zobaa AF. A decentralized and cooperative architecture for optimal voltage regulation in smart grids. *IEEE Transactions on Industrial Electronics*. 2011;58(10):4593–4602.

Chapter 2

Achieving consensus in cooperative and self-organizing sensors network

Sensors networks are considered one of the most promising technologies for decentralized smart grids computing due to their effectiveness in enabling distributed decision-making in networked autonomous systems [1]. In particular, smart sensors may collect and share synchronized grid and environmental variables, adapting the grid controllers parameters via bio-inspired information processing paradigms.

In this context, a challenging issue to address is how to satisfy the strict constraints required for secure and reliable smart grid operation with the limited computing power of the smart sensors and the temporary unavailability of the communication network [2]. Rather than adapting conventional communication architectures to support reliable data-exchange in sensor networks for decentralized smart grids computing, we describe an alternative solution approach, which tries designing both the sensor and the communication network according to a monolithic approach, by explicitly considering the data-centric and event-driven features characterizing the smart sensors.

According to this approach, a sensor network composed of N nodes can be considered as a distributed computing system, which processes all the sensor measurements $x_1, x_2, \ldots, x_N$ in order to compute $f(x_1, x_2, \ldots, x_N)$, where $f(\cdot)$ is a fixed real-valued function and takes proper decisions on the basis of the computed values.

The choice of the function $f(\cdot)$ depends on the application and greatly affects the sensor network design. For example, as proven in [3], if $f(\cdot)$ is invariant to any permutation of the sensor measurements x_i, the network scalability can be sensibly enhanced by deploying in-sensor computing paradigms in the task of exploiting the data-centric feature of the sensor network.

In-sensor computing paradigms could allow for reducing the communication and computational burden of the client/server-based architectures conventionally adopted in smart grids, which allows determining global decisions by collecting, processing, and storing all the sensor measurements x_i by a centralized processing system [4]. Indeed, these paradigms allow distributing the computational intelligence to the entire sensors network, which is able to compute $f(x_1, x_2, \ldots, x_N)$ in a totally distributed way, by only exchanging local information, without the need for deploying a fusion center [5].

For this purpose, a promising research direction is based on the deployment of the so-called distributed consensus protocols, which allow sensors network to reach an agreement on global decisions by only sharing local information, eliminating the need for both explicit point-to-point message passing and routing protocols. Instead, they allow disseminating information throughout the network by updating about the state of each sensor based on a weighted average of its neighboring sensors states. In each iteration, every sensor calculates a local weighted least-squares estimate, which gradually converges to the global maximum-likelihood solution. This unique characteristic enables the sensor network to independently evaluate numerous critical variables characterizing smart grid operation, without relying on a central fusion center to acquire and process all grid measurements. These global variables are then combined with local measurements and processed by each smart sensor to take global decisions aimed at enhancing the smart grid performance.

Sensor networks equipped with distributed consensus protocols have been widely explored in smart grids. In particular, the idea of estimating global grid variables (e.g., average/maximum/minimum system voltage magnitude) by exploiting local coupling among first-order oscillators, which are initialized with the local sensor measurements, has been proposed in [6]. The main idea is to propose a fully decentralized architecture for power quality monitoring, which allow each smart sensor to estimate both the performances of the monitored grid bus, which are directly computed by processing the local measured data, and the global grid performances, which are computed by each smart node by exchanging of information with its neighbors nodes. This allows all the smart sensors to automatically detect power quality anomalies, and the smart grid operator to compute the global power quality indexes by inquiring the state of any sensor without the need of a central fusion center acquiring and processing all the sensor acquisitions.

The same paradigm has been deployed in [7] to coordinate the reactive power generated by renewable power generators in power distribution systems. In particular, the proposed decentralized architecture is based on a network of cooperative smart controllers, which regulate the voltage magnitude of specific smart grid buses by controlling the reactive power generated by the distributed generators in the electrical grids. Thanks to the adoption of distributed consensus protocols, the voltage controllers can infer the main variables characterizing the actual grid operation, compute the actual value of the objective function describing the voltage regulation objectives, and identify in a totally decentralized way the proper control actions aimed at minimizing this function. This decentralized voltage regulation strategy exhibits several advantages over conventional hierarchical control systems, such as lower network bandwidth, higher flexibility, and higher resiliency to communication faults.

These benefits have been confirmed in [8], which proposed the deployment of a network of cooperative phasor measurement units (PMUs) for decentralizing the processing and synchronization functions of wide area monitoring systems. In particular, distributed consensus protocols are here applied in the task of allowing the PMUs to reach time synchronization by adapting their clocks according to proper coupling strategies, without the need for any cluster header, since every PMU is a header to emit synchronization signals to other PMUs. When the PMUs network synchronize, the PMUs clocks are locked to a common phase.

Based on these results, Ref. [9] contributed to the development of a decentralized computing paradigm for solving economic dispatch problems. This computing paradigm is inspired by the mathematics of populations of mutually coupled oscillators, which allow achieving self-synchronization within a distributed control network composed of cooperative and interactive controllers, each responsible for monitoring a specific electric bus. Equipping this controllers network with consensus protocols enables the controllers to perform all the fundamental operations required for solving the economic dispatch problem according to a decentralized and non-hierarchical paradigm.

The integration of fuzzy inference systems and decentralized consensus protocols has been proposed in [10] for optimal voltage regulation in grid-connected photovoltaic systems. The main idea is to decentralize the fuzzy logic-based operators required to compute the optimal set-points of primary voltage controllers on a network of coupled dynamic systems, whose states are updated by proper coupling strategies. This feature allows the dynamic systems to reach a consensus on fuzzy numbers describing the grid performances, hence, adapting the reactive power flows generated by photovoltaic systems.

The role of decentralized consensus protocols in overhead line thermal monitoring has been explored in [11], which proposed a self-organizing sensors network deployed along the line route for decentralizing the entire set of computing and synchronization functions required for identifying the critical span location (i.e., those characterized by the worst heating exchange conditions) and computing the real load-capability of the overhead line in function of the actual conductor thermal state, and the measured environmental conditions. In particular, the smart sensors interacts in order to assess the weather variables and the conductor temperature according to a common virtual clock; verifies the consistency of the measured variables in order to detect sensor faults or data outliers; identifies the line thermal parameters; and dynamically computes the load capability curve of the overhead line. Hence, each smart sensor can compare the local measured variables with the corresponding maximum, minimum, and average values measured along the entire line route, reacting if the local measured variables deviate from fixed confidence intervals. This decentralized bad-data detection feature improves the reliability of the overall computing process.

All these successful results have been obtained by deploying a network of cooperative smart sensors/controllers, whose architecture is based on the following main elements:

1. a set of local transducers measuring both environmental and electrical variables (i.e., wind speed, conductor temperature, voltage magnitude, active and reactive power flow);
2. a first-order dynamical system (i.e., dynamic agent) whose state is initialized by the local measurements and evolves in function of the states of nearby nodes according to decentralized consensus protocols;
3. a short-range communication system, which allows the sharing of the dynamic agent states among the network nodes;

4. local detector/estimator/optimizer aimed at processing the local variables and the dynamic agent states in order to take proper control/regulation actions.

The decentralized consensus protocols adopted for coupling the dynamic agents is based on a challenging idea, originating from [4,12], that is, based on the mathematics of populations of mutually coupled oscillators, where the self-synchronization of the sensors/controllers network is ensured without the need for a fusion center, but only by a proper coupling of the oscillators.

More details about the mathematical backbone supporting this technique are presented and discussed in this chapter, which introduces the mathematical background of consensus protocols and analyzes the conditions that guarantee the network synchronization and the corresponding convergence ratio, and the effects of noise and data-latency on network synchronization.

2.1 Mathematical preliminaries

Let's first consider the case in which each sensor acquires a single variable. In this case, the dynamic agent associated to the ith sensor node evolves according to the following first-order ordinary differential equation [4]:

$$\begin{aligned} \dot{\theta}_i(t) &= \omega_i + \frac{K}{c_i}\sum_{j=1}^{N} a_{ij} f(\theta_j(t) - \theta_i(t)) + v_i(t) \\ \theta_i(0) &= \theta_{i0} \end{aligned} \tag{2.1}$$

where

1. $\theta_i(t)$ is the state of the ith dynamic agent at time t, which could be randomly initialized to a value θ_{i0}, namely $\theta_i(0) = \theta_{i0}$;
2. ω_i is a scalar value, which is related to the variable acquired by the ith sensor, namely $\omega_i = g(x_i)$;
3. $f(\cdot)$ is the dynamic agent coupling function, which could be a linear or nonlinear, odd function, such that

$$\left. \frac{df(x)}{dx} \right|_{x=0} = 1 \tag{2.2}$$

4. the system gain K is a positive constant describing the coupling strength;
5. c_i is a positive parameter describing the capacity of the ith dynamic agent to change its state in the function of the signals received from its neighbor sensor nodes;
6. the parameters $a_{ij} = a_{ji}$ describe the communication topology of the sensors network, namely:

$$a_{ij} = \begin{cases} 1 & \text{if there is a logical connection between the } i\text{th and the } j\text{th node} \\ 0 & \text{otherwise} \end{cases} \tag{2.3}$$

7. $v_i(t)$ describes the additive noise term.

The dynamic agents coupling model defined in (2.1) can be generalized for sensors measuring multiple variables as follows:

$$\begin{aligned} \dot{\theta}_i(t) &= \omega_i + KQ_i^{-1}\sum_{j=1}^{N} a_{ij} f(\theta_j(t) - \theta_i(t)) + v(t) \\ \theta_i(0) &= \theta_{i0} \end{aligned} \tag{2.4}$$

Where

1. L is the number of variables sensed by each sensor;
2. $\theta_i(t)$ is the L-dimension state vector of the ith dynamic agent at time t, which could be randomly initialized to a random vector $\dot{\theta}_{i0}$;
3. ω_i is a L-dimension vector, which is related to the L variables acquired by the ith sensor;
4. Q_i is an invertible L-order-squared matrix describing the observation model.

The dynamic agents coupling model defined in (2.1) and (2.4) for single and multiple measured variables, respectively, can be recasted in a more compact form by using graph theory [12]. The main idea is to model the communication network supporting the sensors data-exchange by an oriented graph $\mathcal{G}$, which includes N vertices, representing the sensors, and N_E edges, representing the communication links between the sensors. The topology of $\mathcal{G}$ is entirely described by its incidence matrix B, which is an NxN_E matrix defined as

$$B_{ij} = \begin{cases} 1 & \text{if the } j\text{th edge enters the } i\text{th vertex} \\ -1 & \text{if the } j\text{th edge leaves the } i\text{th vertex} \\ 0 & \text{otherwise} \end{cases} \tag{2.5}$$

By using this definition, it follows that

$$1_N^T B = 0_{N_E}^T \tag{2.6}$$

where 1_N and 0_{NE} are N and N_E-dimension vectors composed of all ones and zeros, respectively.

The definition of the incidence matrix B allows recasting the coupling model (2.1) as follows:

$$\begin{aligned} \dot{\theta}(t) &= \omega - KD_c^{-1} B D_A f(B^T \theta(t)) \\ \theta(0) &= \theta_0 \end{aligned} \tag{2.7}$$

where

1. $\theta(t) = [\theta_1(t), \ldots, \theta_N(t)]^T$;
2. $\theta_0 = [\theta_{10}, \ldots, \theta_{10}]^T$;
3. D_c is the diagonal matrix of dimension N, whose diagonal elements are c_i;
4. D_A is the diagonal matrix of dimension N_E, whose diagonal elements are a_{ij};
5. $f(x)$ is a vector function defined as:

$$f(x)_k = f(x_k) \ \forall k = [1, N] \tag{2.8}$$

The same vector formalism can be adopted to recast the dynamic agents coupling model (2.4) as follows:

$$\begin{aligned} \dot{\theta}(t) &= \omega - K D_Q{}^{-1} P^T (I_L \otimes B D_A) f[(I_L \otimes B^T) P \theta(t)] \\ \theta(0) &= \theta_0 \end{aligned} \tag{2.9}$$

Where $\otimes$ represents the Kronecker product, and

$$\begin{aligned} \theta(t) &= [\theta_1^T(t), ..., \theta_N^T(t)] \\ \theta(0) &= [\theta_1^T(0), ..., \theta_N^T(0)] \\ \omega &= [\omega_1^T, ..., \omega_N^T] \\ D_Q &= \begin{bmatrix} Q_1 & 0 & ... & 0 \\ 0 & Q_2 & 0 & 0 \\ ... & ... & ... & ... \\ 0 & 0 & ... & Q_N \end{bmatrix} \end{aligned} \tag{2.10}$$

Moreover, the $LNxLN$ matrix P is defined as

$$P_{ij} = \begin{cases} 1 & \text{if } j = ((i \, mod \, N) - 1)L + \frac{i}{N} \, mod(NL) \\ 0 & \text{otherwise} \end{cases} \tag{2.11}$$

The use of this permutation matrix in (2.9) allows each vector $x = [x_1^T, ..., x_N^T]^T$, with $x_i = [x_i^{(1)}, ..., x_i^{(L)}]^T$, to be projected into the vector $\bar{x} = Px$, such that $\bar{x} = [\bar{x}_1^T, ..., \bar{x}_N^T]$, with $\bar{x}_i = [x_1^{(i)}, ..., x_N^{(i)}]^T$.

The solution of the system of ordinary differential equations formalized in (2.1) and (2.4) (or, equivalently, in (2.7) and (2.9)) for single and multiple variables acquisition, respectively, allows each sensor to compute at each time t the current value of both the dynamic agent state $\theta_i(t)$ and its derivative $\dot{\theta}_i(t)$, and to perform proper decisions or estimations by processing these values. In this context, we say that the sensor networks reach a global consensus if, after an initial transient time period $[0, t_s]$, all dynamic agent evolve with the same state, namely

$$\theta_i(t) = \theta_j(t) \;\; \forall t \geq t_s \; \forall i, j = [1, N] \tag{2.12}$$

or with the same state derivative, namely

$$\dot{\theta}_i(t) = \dot{\theta}_j(t) \;\; \forall t \geq t_s \; \forall i, j = [1, N] \tag{2.13}$$

Among these two definitions, in this book, we refer to the second one, namely [4].

Definition 1. *The dynamic agents network coupled by (2.1) (or (2.4)) is said to be synchronized if the derivative of all the state variables $\dot{\theta}(t)$ asymptotically converge to the same solution $\dot{\theta}^*(t)$, namely*

$$\lim_{t \to \infty} \|\dot{\theta}_i(t) - \dot{\theta}^*(t)\| = 0, \;\; i = 1.2, ..., N \tag{2.14}$$

where $\|.\|$ indicates the norm operator. The dynamic network is said to be globally asymptotically stable if the agents synchronize for any initial conditions $\theta_i(0)$.

The adoption of this consensus paradigm is motivated by its enhanced noise rejection features and by the possibility of driving the dynamic agents states to converge to different functions, which could be useful in solving many classes of detection and pattern recognition problems [4,13].

Starting from Definition 1, some interesting features characterizing consensus protocols can be inferred. In particular, if all the dynamic agents converge to a synchronized state, which is globally asymptotically stable, then this equilibrium state is unique and it can be expressed in a closed form by manipulating the differential equations (2.1) and (2.4). Indeed, by left-multiplying equation (2.7) by the vector $c^T = 1_N^T D_c$, and considering that $f(\cdot)$ is an odd function, and $1_N^T B = 0_{N_E}^T$, we get:

$$c^T \dot{\theta}(t) = c^T \omega - K 1_N^T B D_A f(B^T \theta) = c^T \omega \qquad (2.15)$$

Consequently, if the dynamic agents network synchronizes, the common synchronization value can be expressed as

$$\dot{\theta}^*(t) = \omega^* = \frac{c^T \omega}{1_N^T c} = \frac{\sum_{i=1}^N c_i \omega_i}{\sum_{i=1}^N c_i} \qquad (2.16)$$

and we say that the dynamic agents network has reached a consensus on ω^*, which is the weighted average of the variables sensed by all the sensors. This computing paradigm is an instance of decentralized average consensus protocols.

Example 1 (Self-synchronization of dynamic agents observing scalar variables). *Let us consider a self-organizing sensors network composed of four nodes, which acquire local voltage magnitudes and interact over a ring communication topology. The corresponding dynamic agents evolve according to the coupled model (2.1), by assuming the following assumptions:*

- $K = 1$;
- $c_i = 1 \ \forall i \in [1,4]$;
- *randomly generated initial conditions* $\theta_i(0)$;
- $f(\cdot) = tanh(\cdot)$, *namely the dynamic agent coupling function is the hyperbolic tangent function.*

The corresponding trajectories of the derivatives of the dynamic agents state $\dot{\theta}_i(t)$ *for the observed vector* $\omega = [2.05, 2.01, 1.99, 1.95]$ *have been reported in Figure 2.1.*

By analyzing this figure, it is worth observing as the dynamic agents network quickly reaches a consensus on the average value of the observed variables, which is $2V$.

The same results can be obtained for sensors network measuring multiple variables. In this case, by left-multiplying (2.9) by the matrix $(1_N^T \otimes I_L) D_Q$, and considering the following algebraic expressions:

$$(1_N^T \otimes I_L) P^T (I_L \otimes B D_A) = (I_L \otimes 1_N^T)(I_L \otimes B D_A) = I_L \otimes 1_N^T B D_A = 0_{LxLN_E} \qquad (2.17)$$

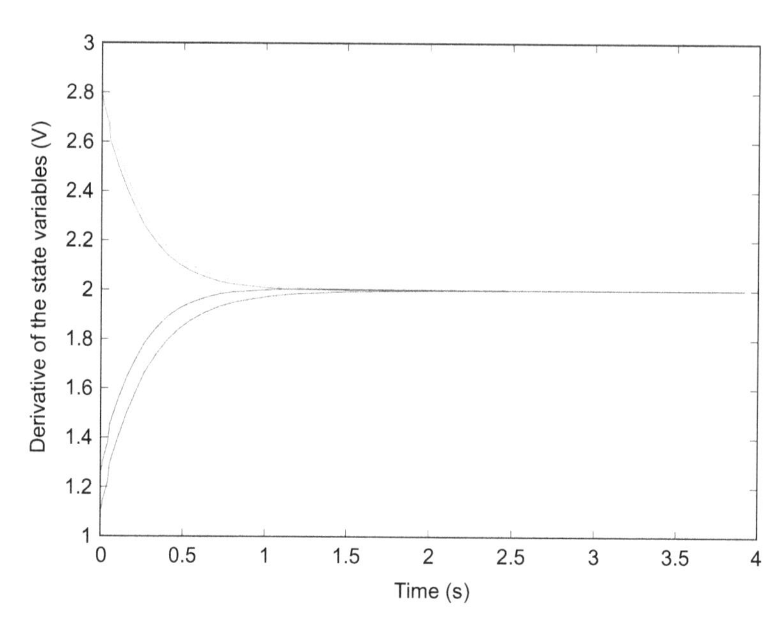

Figure 2.1 Example of self-synchronization of four dynamic agents acquiring a single variable and interacting over a ring topology

we get the following equation:

$$\sum_{i=1}^{N} Q_i \dot{\theta}_i(t) = \sum_{i=1}^{N} Q_i \omega_i \tag{2.18}$$

Thanks to this equation, it is possible to conclude that if the dynamic agents network synchronizes, the common synchronized state is expressed as

$$\dot{\theta}^*(t) = \omega_L^* = \left(\sum_{i=1}^{N} Q_i \right)^{-1} \left(\sum_{i=1}^{N} Q_i \omega_i \right) \tag{2.19}$$

Example 2 (Self-synchronization of dynamic agents observing multiple variables). *Let us consider the same self-organizing sensors network considered in Example 5, but acquiring both local voltage and current magnitudes. In this case, the observed vectors are composed of the voltage and current magnitudes measured at each node, which have been assumed as*

$$\begin{aligned} \omega_1 &= [2.05, 0.10] \\ \omega_2 &= [2.01, 0.25] \\ \omega_3 &= [1.99, 0.15] \\ \omega_4 &= [1.95, 0.3] \end{aligned} \tag{2.20}$$

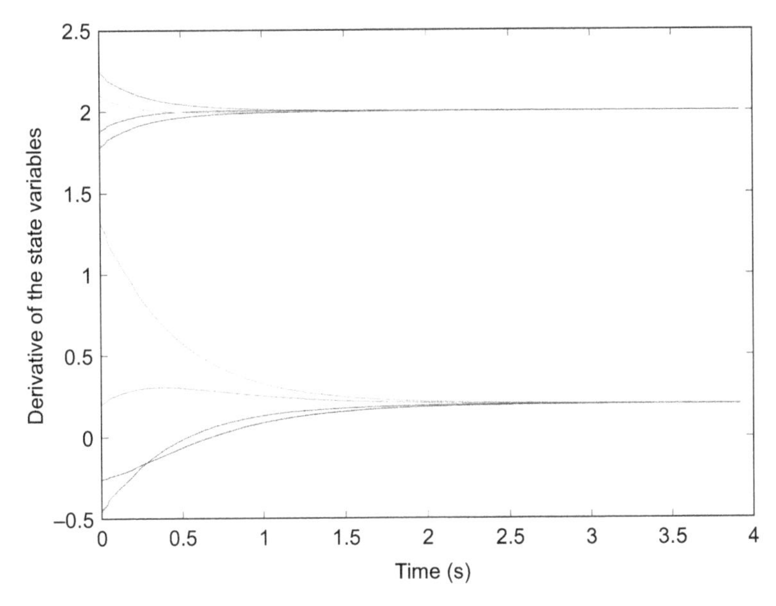

Figure 2.2 Example of self-synchronization of four dynamic agents acquiring two variables and interacting over a ring topology

As confirmed by the trajectories reported in Figure 2.2, also in this case, the dynamic agents evolving according to the vector coupling model (2.4) quickly synchronize on the mean values of the measured local variables, which are 2V and 0.2A.

The spontaneous emergence of synchronization from cooperative dynamic agents has been widely analyzed in the computational intelligence literature and applied in the task of solving heterogeneous estimation and detection problems. One of the most relevant features characterizing these solution paradigms is the "think locally act globally" principle, which allows simple, decentralized agent behaviors to guarantee the reliable synchronization of the overall network, despite local sensors or communication faults, or changes in network topology. This important feature is typically referred as "Self-organizing" network.

2.2 Solving decentralized estimation problems

The decentralized consensus protocols introduced in Section 2.1 allows the sensors network monitoring a physical phenomenon to compute global maximum-likelihood estimations by only exchanging local information about the dynamic agents state, without the need for sending all the measured variables to a fusion center.

In particular, let us assume that the sensors network should estimate the value of an unknown physical parameter ξ by processing the sensor measurements x_i, which can be expressed as

$$x_i = b_i\xi + \eta_i \quad \forall i \in [1,N] \tag{2.21}$$

where $(\eta_1, ..., \eta_N)$ are additive noise terms, which are described by Gaussian random variables. For the sake of simplicity, these random variables have been assumed to be independent and identically distributed with zero mean and variances $(\sigma_1^2, ..., \sigma_N^2)$.

This estimation problem can be solved by exploiting the self-synchronization feature of the dynamic agents network. For this purpose, let us define ω_i and c_i as follows:

$$\begin{aligned} \omega_i &= \frac{x_i}{b_i} \\ c_i &= \frac{b_i^2}{\sigma_i^2} \quad \forall i \in [1,N] \end{aligned} \tag{2.22}$$

Hence, according to (2.16), if the dynamic agents network synchronizes, the agents reach a consensus on the following synchronized state:

$$\omega^* = \hat{\xi}_{ML} = \frac{\sum_{i=1}^{N} \frac{b_i x_i}{\sigma_i^2}}{\sum_{i=1}^{N} \frac{b_i^2}{\sigma_i^2}} \tag{2.23}$$

This synchronized state is exactly the maximum-likelihood estimation of the unknown physical parameter ξ [14].

By analyzing (2.23), it is worth noting that the computed estimation is mainly ruled by the most accurate sensors, which are the ones characterized by the smallest standard deviations. That is why the parameters c_i, which regulate the attitude of the dynamic agents to adapt their state according to the variation of their neighbor states, have been parameterized to the sensor accuracy. Indeed, the higher the sensor accuracy, the less the change of the state function of the corresponding dynamic agent, because the sensor measurement can be considered reliable. In this context, it is important to note that the maximum-likelihood estimation of the unknown parameter is obtained without sharing the sensors accuracy (e.g., the signal-to-noise ratio of each sensor) along the sensors network.

These results can be generalized to the estimation of an unknown L-dimension parameters vector ξ. In this case, let's consider a linear observation model, assuming that each sensor acquires the following M-dimension observation vector:

$$x_i = A_i\xi + \eta_i \tag{2.24}$$

where A_i is the observation matrix, whose dimension is MxL, and η_i is the additive Gaussian noise vector, which have been assumed statistically independent with zero mean and covariance matrix C_i.

For the sake of simplicity, we assume that each sensor can compute the local estimation of the unknown parameter vector ξ, by processing its measurements.

According to this hypothesis, it follows that $M \geq L$ and the observation matrix A_i should be full column rank. Hence, by making the following assumptions in (2.9):

$$
\begin{aligned}
Q_i &= A_i^H C_i^{-1} A_i, \\
\omega_i &= (A_i^H C_i^{-1} A_i)^{-1} A_i^H C_i^{-1} x_i
\end{aligned}
\tag{2.25}
$$

It follows that if the dynamic agents network synchronizes, then, according to (2.19), the agents reach a consensus on the following synchronized state:

$$
\omega_L^* = \hat{\xi}_{ML} = \left(\sum_{i=1}^{N} A_i^H C_i^{-1} A_i \right)^{-1} \left(\sum_{i=1}^{N} A_i^H C_i^{-1} x_i \right) \tag{2.26}
$$

which, again, it is exactly the maximum-likelihood estimation of the unknown vector ξ [14].

Also in this case, it is important to emphasize some intrinsic benefits deriving by the application of the dynamic agents network in decentralized estimation of multiple physical parameters. Indeed, the solution to this estimation problem by a conventional hierarchical centralized computing approach requires the collection of all the observation vectors x_i, all the observation matrices A_i, and all the covariance matrices C_i. On the contrary, by using the described decentralized approach, all the dynamic agents, if the network synchronizes, reach a consensus on the optimal maximum-likelihood estimation of the unknown parameters vector, without the need for sharing all the data on the network, but only by exchanging their state vectors $\theta_i(t)$ with their neighbors.

These benefits may be also obtained by deploying the described sensors network for solving further classes of problems. Indeed, by considering that ω_i can be fixed according to the local sensor measurements, namely $\omega_i = g(x_i)$, and according to (2.16), it can be concluded that the dynamic agents network can reach a consensus on any function of the measured data $f(x_1, \ldots, x_N)$ that can be expressed as

$$
f(x_1, \ldots, x_N) = \frac{\sum_{i=1}^{N} c_i g(x_i)}{\sum_{i=1}^{N} c_i} \tag{2.27}
$$

As we will see in the next chapters, many fundamental smart grids operation problems can be formalized according to this computing paradigm and solved by a dynamic agents network according to a fully decentralized scheme. On the other hand, it is worth noting that dynamic agents should deploy an iterative algorithm to synchronize and reach a consensus on general functions. Hence, the conditions guaranteeing the network synchronization and the corresponding convergence ratio are two fundamental issues that should be analyzed in order to enabling the effective deployment of self-organizing dynamic agents in decentralized smart grids computing.

2.3 Conditions for network synchronization

The conditions guaranteeing the dynamic agents synchronization, as far as the corresponding convergence time, are strictly dependent on the network topology, and, in particular, by its Laplacian L and weighted Laplacian L_A, which are defined as

$$\begin{aligned} L &= BB^T \\ L_A &= BD_AB^T \end{aligned} \tag{2.28}$$

where B is the incidence matrix of the oriented graph $\mathcal{G}$ describing the network topology and D_A is a diagonal matrix of dimensions $N_E x N_E$, whose diagonal elements are the edge weights, namely

$$D_A = \begin{bmatrix} e_1 & 0 & \dots & 0 \\ 0 & e_2 & \dots & 0 \\ \dots & \dots & \dots & \dots \\ 0 & 0 & \dots & e_{N_E} \end{bmatrix} \tag{2.29}$$

where

$$e_k = a_{ij} \quad \forall k \in [1, N_E] \, \forall i, j \in [1, N] \tag{2.30}$$

Observe that L is independent of the graph orientation, and both L and L_A are characterized by the following proprieties [15]:

- they are positive semi-definite;
- their smallest eigenvalue is null and its algebraic multiplicity depends by the number of connected graph components (e.g., 1 for connected graphs);
- if the graph is connected then $rank(L) = rank(L_A) = N - 1$, hence, the eigenvector associated to the smallest eigenvalue is 1_N;
- their second smallest eigenvalue $\lambda_2(L)$ and $\lambda_2(L_A)$ directly influences the algebraic graph connectivity [15].

On the basis of these proprieties, the following theorem allows determining the necessary and sufficient conditions guaranteeing the stable synchronization of the dynamic agents network observing scalar variables [4].

Theorem 1. *Let us consider a network of dynamic agents interacting over a communication network described by an oriented graph $\mathcal{G}$, evolving according to the coupled model defined in (2.1) and satisfying the following hypothesis:*

1. *$\mathcal{G}$ is connected;*
2. *$f(\cdot)$ is continuously differentiable, odd, not-decreasing, and asymptotically convex or concave, namely*

$$\exists x \in R : sign\left(\frac{df^2}{dx^2}(x)\right) = sign\left(\frac{df^2}{dx^2}(\bar{x})\right), \forall x \geq \bar{x}; \tag{2.31}$$

3. *$a_{ij} \geq 0 \; \forall i, j \in [1, N]$;*
4. *$c_i \geq 0 \; \forall i \in [1, N]$.*

Then, there exist two unique and positive thresholds, K_L and K_U, with $K_L < K_U$, such that

1. *$\forall K \geq K_U$ it exists in a synchronized state that is globally asymptotically stable;*
2. *$\forall K < K_L$ it does not exist in a synchronized state.*

Moreover, the corresponding lower and upper bounds of these thresholds may be computed as

1. *if $f(\cdot)$ is unbounded, then*

$$\begin{aligned} K_L &= 0 \\ K_U &= 0 \end{aligned} \tag{2.32}$$

2. *If $f(\cdot)$ is bounded, then*

$$\begin{aligned} K_L &\geq \frac{\|D_c \Delta\omega\|_\infty}{f_{max} d_{max}} \\ K_U &\leq \frac{2\|D_c \Delta\omega\|_2}{f_{max} \lambda_2(L_A)} \end{aligned} \tag{2.33}$$

where

- $\Delta\omega = [\omega_1 - \omega^*, \ldots, \omega_N - \omega^*]$;
- $f_{\max} = \lim_{x \to +\infty} f(x)$;
- $d_{\max} = \max_i \sum_{j=1}^{N} a_{ij}$;
- *$\lambda_2(L_A)$ is the algebraic connectivity of $\mathcal{G}$, namely the second smallest eigenvalue of the weighted Laplacian L_A.*

This fundamental theorem allows us to design the sensors network and selecting the system gain K in order to guarantee the dynamic agents synchronization. Indeed, conditions (2.32) and (2.35) allow determining the minimum value of K, which guarantee the dynamic agents to reach a consensus. Hence, by simply varying a single parameter over specified ranges, which only depend on the communication network topology, it is possible to control the network synchronization. This important feature, which requires selecting a bounded coupling function satisfying Theorem 1 hypothesis, can be extremely useful when the cooperative sensors network is deployed in the task of solving specific spatial clustering problems [12]. On the other hand, the adoption of a linear coupling function (i.e., $f(x) = x$) does not allow controlling the network synchronization by changing the system gain, since, as for any unbounded coupling functions, the dynamic agents network synchronizes for any $K > 0$, and the only condition avoiding the network synchronization is $K = 0$.

Another relevant information that can be inferred from this theorem concerns with the convergence ratio of the synchronization process. In particular, if a synchronized state exists (i.e., $K \geq K_U$), and $c_i = 1\ \forall i \in [1, N]$, then the dynamic agents reach the consensus on this state with a convergence ratio that depends on $K\lambda_2(L_A)$. This important result emphasizes the potential role played by the system gain K and the algebraic connectivity of the graph topology in enhancing the convergence performance of the agents network, reducing the number of synchronization messages that the dynamic agents should exchange, and the sensors energy consumption, which is an important issue to address if the sensors are powered by batteries.

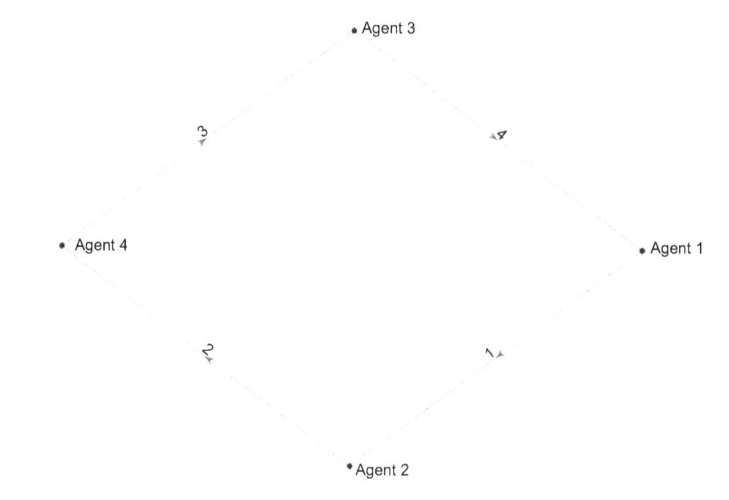

Figure 2.3 Example of four dynamic agents interacting over a ring topology

Example 3 (Effect of K on self-synchronization). *Let us consider the self-organizing sensors network considered in Example, which is composed of four sensors interacting over the graph reported in Figure 2.3, which is characterized by the following incidence matrix:*

$$B = \begin{bmatrix} -1 & 0 & 0 & 1 \\ 1 & -1 & 0 & 0 \\ 0 & 0 & 1 & -1 \\ 0 & 1 & -1 & 0. \end{bmatrix} \tag{2.34}$$

Then by considering that

1. $\lambda_2(L_A) = 2;$
2. $f_{max} = 1;$
3. $\Delta\omega = [0.05, 0.01, -0.01, -0.05]$
4. $D_C = \begin{bmatrix} 1 & 0 & 0 & 0 \\ 0 & 1 & 0 & 0 \\ 0 & 0 & 1 & 0 \\ 0 & 0 & 0 & 1 \end{bmatrix}$

The application of (2.35) allows determining the bounds of K_U and K_L, namely $K_U \leq 0.0721$ and $K_L \geq 0.025$.

Hence, if the system gain K is lower than 0.025, then the dynamic agents evolving, according to the coupled model (2.1), does not reach a common synchronization state, as confirmed in Figure 2.4, which reports the trajectories of the state variables derivatives for $K = 0.01$.

Finally, it is worth noting that Theorem 1 can be applied also for time-varying graph topology, which can be useful in modeling communication channels affected

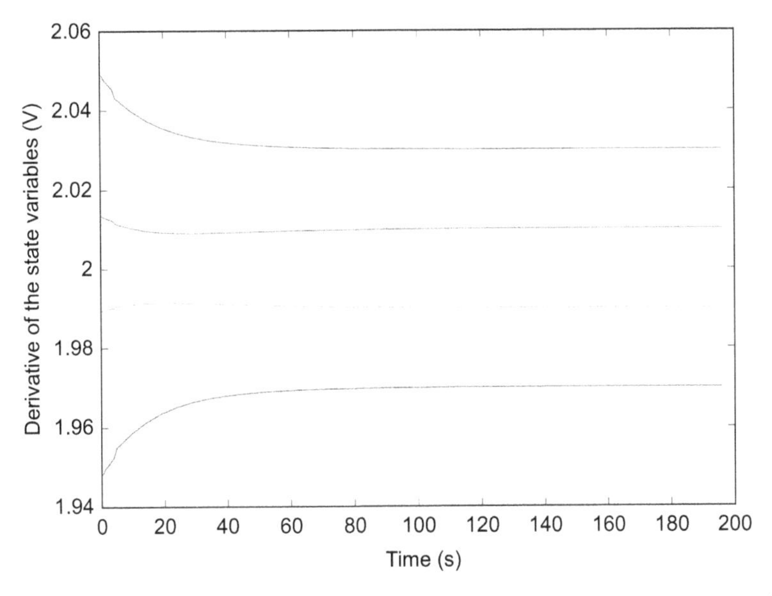

Figure 2.4 Example of four dynamic agents not reaching a global consensus due to $K < K_L$

by fading. For this purpose, some coefficients a_{ij} are represented by positive random variables, and the incidence matrix becomes a random matrix. In this context, if the communication network topology obtained by only considering the deterministic a_{ij} is connected, then the hypothesis of Theorem 1 is satisfied, and the dynamic agents can reach a consensus according to conditions (2.35). Hence, if the deterministic portion of the graph is connected, the capability of reaching a consensus by network synchronisation does not depend by the randomness of a_{ij}. While the values assumed by these coefficients influence the algebraic connectivity of the graph, namely $\lambda_2(L_A)$, and consequently the convergence ratio to the synchronization state.

Although these results have been obtained with reference to dynamic agents network observing a single (scalar) variable, they can be properly generalized for multiple variables observation. In particular, by assuming the same hypothesis of Theorem 1, the following conditions stand [4]:

1. If $f(\cdot)$ is unbounded then $K_L = K_U = 0$.
2. If $f(\cdot)$ is bounded, the bounds of the critical synchronization thresholds are

$$\begin{aligned} K_L &\geq \frac{\max_i \|\Delta\bar{\omega}_i\|_\infty}{f_{\max} d_{\max}} \\ K_U &\leq \frac{\max_i \|\Delta\bar{\omega}_i\|_2}{f_{\max} \lambda_2(L_A)} \end{aligned} \tag{2.35}$$

where $\Delta\bar{\omega} = [\Delta\bar{\omega}_1^T, \ldots, \Delta\bar{\omega}_L^T] = PD_Q(\omega - 1_N \otimes \bar{\omega}_L^*)$.

Hence, all the argumentation discussed for the scalar observation can be properly extended to the vector observation case.

2.4 Effect of network topology on self-synchronization

As observed in Section 2.3, if a convergence state exists, then the dynamic agents network self-synchronizes to this state with a convergence ratio that depends on the algebraic connectivity of the communication network graph, i.e., $\lambda_2(L_A)$. Hence, for a fixed $K \geq K_U$, the number of iterations required for dynamic agents synchronization strictly depends on the communication network topology, and, more specifically, on the number of neighbors linked to each agent, which is a topological feature characterizing the graph $\mathscr{G}$ known as the node degree. In particular, if $\mathscr{G}$ is a regular graph, namely all their nodes are characterized by the same degree, then the convergence ratio can be expressed in closed form, while, in general, the algebraic connectivity of $\mathscr{G}$ can be bounded for $N \gg 1$ as follows [16]:

$$\lambda_2 \geq 2\left(1 - \cos\frac{\pi}{N}\right)\delta \approx \frac{\pi^2}{N^2}\delta \tag{2.36}$$

where δ is the minimum node degree. Consequently, for a fixed δ, if the dynamic agents number increases, then λ_2 decreases as $1/N^2$, and, according to Theorem 1, the system gain K should be increased as N^2 in order to guarantee the existence of a synchronized state that is globally asymptotically stable.

Different proprieties can be defined for irregular graphs. For example, small-world communication networks, in which most agents are linked by a small hops number, are characterized by larger algebraic connectivity compared to regular networks, hence exhibiting higher convergence ratios. Moreover, if the communication network is described by a scale-free graph, which is obtained by starting from an initial set of sensors that is iteratively increased according to proper growth policies, then the limit of the average value of the graph algebraic connectivity tends to a constant value for sensors nodes going to infinity [4,17]. Thanks to this important theoretical result, we can conclude that dynamic agents interacting over scale-free communication topologies are highly scalable and fault tolerant, since for $K > K_U$, and a sufficiently large number of sensors, the addition/removal of a limited number of sensors does not affect the self-synchronization of the dynamic agents, and, consequently, their estimation and detection features.

2.5 Effects of data uncertainties on self-synchronization

The decentralized consensus protocols described in (2.1) or (2.4) allow the dynamic agents to reach a consensus on global variables, enabling them to cooperate similarly to a population of mutually coupled oscillators. For this purpose, the main tasks that each dynamic agent periodically performs are

1. measure the local variables by querying the available sensors;
2. data filtering and processing;
3. collect the agents state of its neighbors;
4. update the state of its dynamic agent according to (2.1) or (2.4);
5. share the updated agent state with its neighbors.

All these tasks may be affected by heterogeneous uncertainty sources, which perturb the capability of dynamic agents network in reaching a consensus, such as the measurement noise affecting the sensors acquisitions, which influence the determinism of ω_i, and the communication errors, which affect the agents coupling and can be modeled by integrating the additive noise terms $v_i(t)$ and $v(t)$ in (2.1) and (2.4), respectively.

These uncertainties are derived from several sources, such as measurement uncertainties induced by the voltage/current transducers, data processing errors, and non-ideal communication channels. Moreover, dynamic agents-based decentralized estimation requires the deployment of complex mathematical operators, which include linear regression, and matrix inversion. The deployment of these mathematical operators on microcontroller-based units requires large computing times, introducing accumulation errors and delays on the overall computing chain, which could compromise the estimation accuracy, especially in the presence of random clock instabilities.

Other relevant uncertainty sources are related to the residual errors of the synchronized systems, which are induced by the accuracy of the synchronization algorithms, and the clock drift [18].

Finally, a self-organizing sensors network typically requires the availability of timestamps at the medium access channel level, which are often generated at the application level. Hence, these timestamps may also be affected by communication latencies [19].

The main effect of all these uncertainties in solving decentralized estimation problems is that the network synchronization to the global maximum-likelihood estimation of the unknown physical parameters is no longer guaranteed in the presence of unbounded observation noise. Indeed, for any fixed value of the system gain K, the probability $K_L > K$ is greater than zero, hence, hindering the dynamic agents to reach a global consensus. In particular, if we model the observation noises by random Gaussian variables, then the probability of not reaching a consensus can be made arbitrarily small by raising the system gain K [4], as shown in the following example.

Example 4 (Decentralized estimation problem in the presence of observation noise). *Let us consider a network of 32 sensors interacting over a 6-degree communication graph, namely each sensor shares the state of its dynamic agents with other 6 neighbors. The function describing the dynamic agents coupling is $f(x) = tanh(x)$, which is a nonlinear bounded function. The sensors network aims at estimating a vector of three unknown parameters by using the following linear observation model:*

$$x_i = A_i\xi + \eta_i \tag{2.37}$$

where the 6x3 *observation matrices A_i have been assumed composed of statistically independent Gaussian random variables with zero mean and unit variance, and the observation noise is modeled by white Gaussian noise with zero mean and unit variance. The real value of the unknown parameters vector is $\xi = [5, 1, 15]^T$.*

The solution of the estimation problem described in Example 4 by the decentralized model formalized in (2.4) allows obtaining the derivative state trajectories

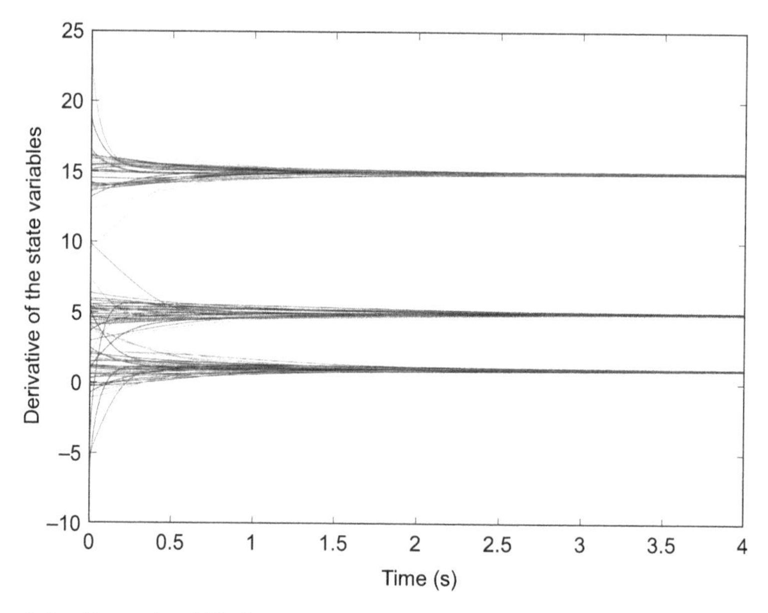

Figure 2.5 Example of 32 dynamic agents reaching a global consensus on the maximum-likelihood estimation of a three-dimensional unknown vector

reported in Figure 2.5. In the same figure, the obtained estimations have been benchmarked with the solutions obtained by deploying a conventional centralized estimation scheme, which collects the correct values of all the sensor observations and the mixing matrices. The analysis of this comparative analysis confirmed the reliable convergence of all the dynamic agents to the global maximum-likelihood estimation of the physical parameters.

In order to assess the estimation robustness of the decentralized solution scheme, the same estimation problem has been solved for different random instances of the mixing matrices, initial conditions and observation noises, by using an increasing numbers of sensors with a fixed neighbors number (i.e., six-degree regular graph). The obtained results have been summarized in Figure 2.6, which reports the average values of the estimation variance characterizing the self-organizing dynamic agents in function of the sensors number. By analyzing this figure, it is worth noting the robustness and the estimation accuracy characterizing the decentralized scheme, which allows computing reliable estimations of the unknown parameters that are very close to the solutions computed by the conventional centralized estimation scheme.

Moreover, considering that the estimation is computed by only sharing local information, which does not affected by N since the dynamic agents interacts on a fixed graph degree, and the average estimation variances decay as the sensors number increase, it is possible to assess the good scalability level of the decentralized estimation scheme.

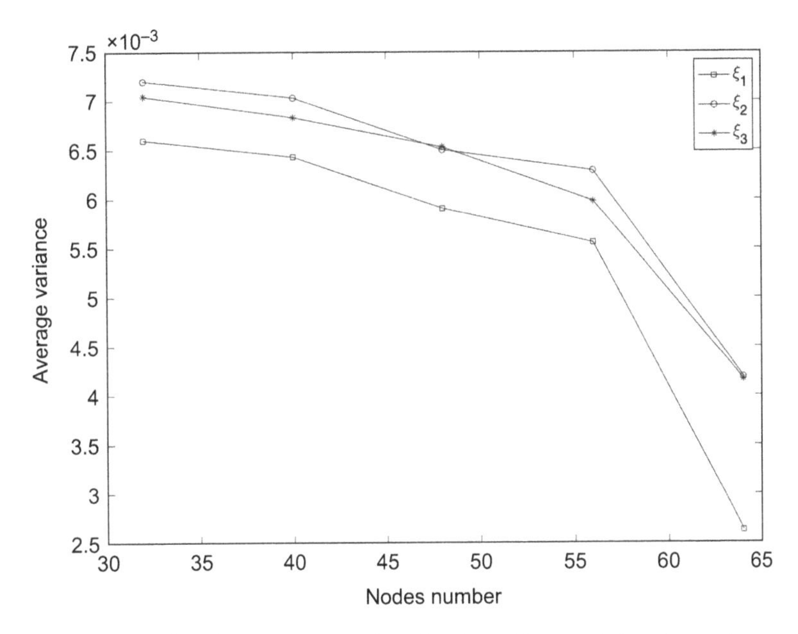

Figure 2.6 Example of parameters estimation by self-organizing dynamic agents: effect of the sensors number on the mean estimation variance

2.6 Effects of communication latency on self-synchronization

The propagation delays of the communication channel depend on many random phenomena, which are related to the particular architecture deployed for supporting data-exchange in the sensors network [12]. The effects of these delays can be modeled by adding random time offsets in the messages that the dynamic agents periodically exchange in order to share their states. In this context, the self-synchronization capability of the dynamic agents network evolving according to the coupling model (2.1) or (2.4) strictly depends on the regularity of the network graph $\mathcal{G}$, and, in particular, of the algebraic properties of its Laplacian, as stated in the next theorem, which allows determining the necessary and sufficient conditions guaranteeing the dynamics agent network measuring a scalar variable to reach a consensus on common synchronization states in the presence of asymmetric propagation delays [12].

Theorem 2. *Let us consider a network of dynamic agents interacting over a communication network described by an oriented graph $\mathcal{G}$, evolving according to the coupled model defined in (2.1), and satisfying the following hypothesis:*

1. $K > 0$;
2. $c_i > 0 \ \forall i \in [1, N]$;
3. $a_{ij} \geq 0 \ \forall i, j \in [1, N]$;
4. *the finite propagation delays τ_{ij} are constant;*

5. *$f(\cdot)$ is a continuously, differentiable, and bounded function;*
6. *the initial conditions θ_{i0} $\forall i \in [1,N]$ are taken in the interval $[-\tau,0]$, where τ is the maximum propagation delay (i.e., $\tau = max_{i\neq j}\tau_{ij}$).*

Then the dynamic agents network globally synchronizes for any instance of the propagation delays, if and only if the graph $\mathcal{G}$ is quasi-strongly connected, namely it has a center (root) node from which any other node is reachable. The corresponding synchronization state is

$$\omega^* = \frac{\sum_{i=1}^{N} \gamma_i c_i \omega_i}{\sum_{i=1}^{N} \gamma_i c_i + K \sum_{i=1}^{N} \sum_{j\in N_i} \gamma_i a_{ij} \tau_{ij}} \tag{2.38}$$

where $[\gamma_1, ..., \gamma_N]$ is the left eigenvector of the Laplacian L corresponding to the null eigenvalue.

These important results can be generalized for the case of dynamics agent network measuring a set of variables, as stated in the following theorem [20].

Theorem 3. *Let us consider a network of dynamic agents interacting over a communication network described by an oriented graph $\mathcal{G}$, evolving according to the coupled model defined in (2.4), and satisfying the following hypothesis:*

1. $c_i > 0$ $\forall i \in [1,N]$;
2. $a_{ij} \geq 0$ $\forall i, j \in [1,N]$;
3. *Q_i are definite positive $\forall i \in [1,N]$.*

Then the dynamic agents network globally synchronize for any instance of the propagation delays, if and only if the graph $\mathcal{G}$ is quasi-strongly connected, and the corresponding synchronization state is

$$\omega^* = \left(\sum_{i=1}^{N} \gamma_i Q_i + I_L \otimes \left(K \sum_{i=1}^{N} \sum_{j\in N_i} \gamma_i a_{ij} \tau_{ij} \right) \right)^{-1} \left(\sum_{i=1}^{N} \gamma_i Q_i \omega_i \right) \tag{2.39}$$

These two fundamental theorems allow determining the conditions for the self-synchronization of the dynamic agents, and, if it exists, the corresponding synchronization state, which depends on the graph topology (i.e., a_{ij} and γ_i) and the propagation delays (i.e., τ_{ij}). Hence, if the network graph is quasi-strongly connected and the theorem hypothesis is satisfied, we can argue that data latency does not hinder the dynamic agents to reach a consensus by self-synchronization, but it perturbs the value of the corresponding synchronization state, by introducing a bias, which depends on the graph topology and the propagation delays. In this context, the dynamic agents converge to a synchronized state that is different from the ideal one, which is achievable only in the presence of ideal communication links (i.e., with $\tau_{ij} = 0$). To compensate this difference, it is possible to deploy a two-step estimation process, which requires the dynamic agents to reach a first consensus by assuming a fixed fictitious measurement (i.e., $\omega_i = 1$), hence, synchronizing to

$$\omega^*(1) = \frac{\sum_{i=1}^{N} \gamma_i c_i}{\sum_{i=1}^{N} \gamma_i c_i + K \sum_{i=1}^{N} \sum_{j\in N_i} \gamma_i a_{ij} \tau_{ij}} \tag{2.40}$$

after synchronizing to this state, the dynamic agents evolve again in the task of reaching another consensus by considering the real measured variables ($\omega_i = g(x_i)$), hence, synchronizing to

$$\omega^* = \frac{\sum_{i=1}^{N} \gamma_i c_i \omega_i}{\sum_{i=1}^{N} \gamma_i c_i + K \sum_{i=1}^{N} \sum_{j \in N_i} \gamma_i a_{ij} \tau_{ij}} \tag{2.41}$$

Then, by considering the ratio of these two synchronization states

$$\frac{\omega^*}{\omega^*(1)} = \frac{\sum_{i=1}^{N} \gamma_i c_i \omega_i}{\sum_{i=1}^{N} \gamma_i c_i} \tag{2.42}$$

It is possible to compute the synchronization state by compensating the effects of the propagation delays according to a fully decentralized scheme and without requiring any knowledge or estimation of any communication network parameters [20].

Example 5 (Self-synchronization in the presence of asymmetric propagation delays). *Let us analyze the synchronization of the self-organizing sensors network considered in Example in the presence of the following asymmetric propagation delays (expressed in seconds):*

$$\tau = \begin{bmatrix} 0 & 0.1 & 0.2 & 0 \\ 0.3 & 0 & 0 & 0.4 \\ 0.1 & 0 & 0 & 0.5 \\ 0 & 0.2 & 0.3 & 0 \end{bmatrix} \tag{2.43}$$

The derivative state trajectories of the dynamic agents computed by considering the effects of these propagation delays are reported as continuous lines in Figure 2.7. In the same figure, the dynamic trajectories computed in the ideal case (e.g., neglecting the propagation delays) are reported as dash-dotted lines.

By analyzing this figure, it is worth noting that, as dictated by Theorem 2, the presence of asymmetric propagation delays does not affect the self-synchronization feature of the dynamic agents network, but they influence the common synchronization state, which is now equal to 1.34, and differ from the ideal value 2. Hence, the deployment of the two-step compensation process is required in order to allow the dynamic agents to reach a consensus on the correct synchronization state. Indeed, by synchronizing the dynamic agents on a fixed fictitious measurement (i.e., $\omega_i = 1\ \forall i \in [1,4]$*), the trajectories reported in Figure 2.8, which, as expected, globally converge to* 0.66.

This two-step estimation process should be activated during the sensors network start-up, and in the presence of any variation of the graph topology, or the communication links.

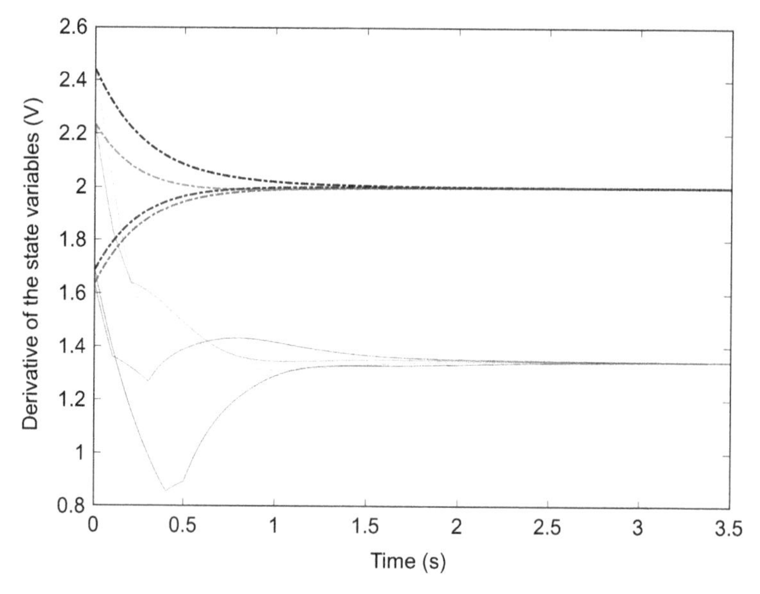

Figure 2.7 Example of self-synchronization of four dynamic agents in the presence of propagation delays, the continuous and dash-dotted lines represent the real and ideal trajectories, respectively

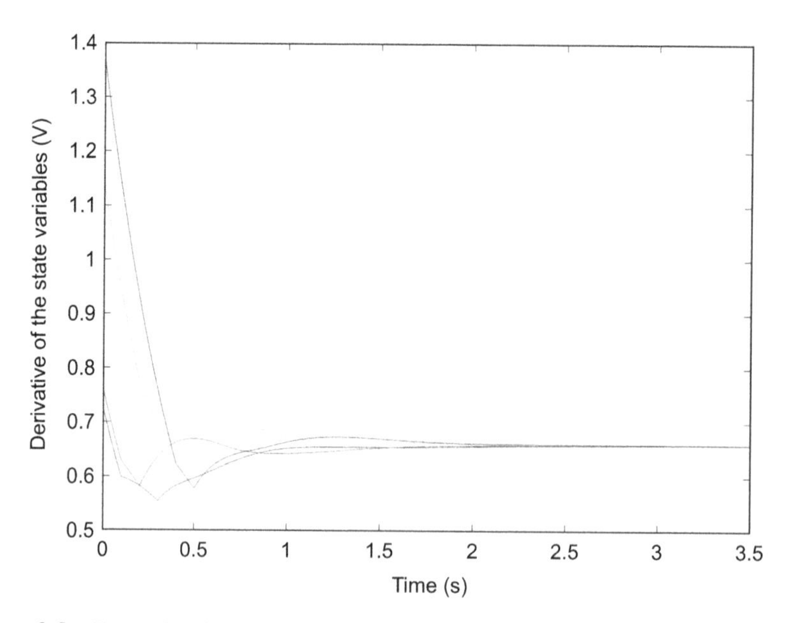

Figure 2.8 Example of self-synchronization of four dynamic agents on fictitious measurement (i.e., $\omega_i = 1 \forall i \in [1,4]$) in the presence of propagation delays

2.7 Reliable computing-based consensus protocols

Distributed consensus protocols enable dynamic agents to reach an agreement on global information or shared variables, facilitating their cooperation. In this context, each agent is tasked with specific tasks to achieve decentralized consensus, including

1. obtaining relevant variables by querying local sensors;
2. processing raw measurement data;
3. receiving information about the state of neighboring agents;
4. computing and transmitting its own state.

These tasks introduce uncertainties stemming from various sources, broadly categorized as measurement uncertainties (e.g., nonideal voltage/current transducers, noisy data, and measurement metrics), data processing errors (e.g., rounding errors), and communication errors.

Additionally, decentralized estimation relying on consensus protocols involves many mathematical operations such as linear regression, matrix inversion, and polynomial fittings. The computational complexity of these operations, when executed using programmable units, may lead to non-negligible computing times. The extent of delay depends on factors such as hardware architecture and clock rate, introducing accumulation errors that can impact the overall estimation performance, especially in the presence of clock instabilities. Synchronized systems also exhibit residual errors induced by the accuracy of synchronization algorithms and clock drift [18].

Moreover, consensus protocols for sensor time synchronization typically employ timestamp generation at the medium access channel level. Given that these algorithms are commonly implemented at the application level, the timestamps generated may be influenced by latencies arising from concurrent execution of the measurement application engaged in critical measurement tasks.

The presence of uncertainties within the network of agents has the potential to significantly jeopardize the accuracy and convergence properties of the decentralized consensus protocols they implement. Consequently, the utilization of advanced tools capable of representing and managing these uncertainties is required for the dependable and efficient deployment of such computing paradigms in realistic operation scenario.

One promising approach to address this challenge is the application of self-validated computing. These models inherently capture the accuracy of computed quantities as part of the computation process, eliminating the need for specific information about the type of uncertainty in the parameters. Among the simplest and most widely used forms of self-validated computing is Interval Analysis (IA), allowing for numerical computations where each quantity is represented by an interval of floating-point numbers without a probability structure [21]. Operations such as addition, subtraction, and multiplication on these intervals are performed in a way that guarantees each computed interval contains the unknown value of the represented quantity.

However, it is worth noting that IA does not offer a foolproof guarantee against the introduction of spurious values into the results of interval operations. Moreover, the conservative nature of IA may render it unsuitable for iterative algorithms. Consequently, the interval solutions generated during the computing process are often considerably wider than the actual range of the corresponding quantities, especially in the presence of computational chains where the interval width could diverge.

To mitigate the excessive conservatism associated with IA, an enhanced technique for interval computation based on affine arithmetic (AA) has been proposed in [22]. Through the use of AA, distributed consensus protocols can be tackled by considering the interdependence of parameter uncertainty and the diversity of uncertainty sources.

2.7.1 Problem formulation

The challenges associated with achieving decentralized consensus in the context of data uncertainty can be represented using an interval differential equation (IDE). An IDE is an extension of an ordinary differential equation that incorporates intervals for initial conditions and/or parameters.

Specifically, when the parameters and/or initial value of the dynamic system (2.1) are not precisely determined but instead fall within a certain interval, the model transforms into a differential inclusion. This means that the differential equation accounts for a range of possible values for the parameters and/or initial value, namely

$$\begin{aligned} \dot{x} = g(x, k_1, \ldots, k_p)\; x(0) \in I_x(0) \\ k_j \in I_{k_j} \subset R\; j = 1, \ldots, p \end{aligned} \tag{2.44}$$

where the initial condition belongs to the m-dimensional hypercube $I_x(0)$, and the value of each uncertain parameters k_j is unknown but bounded in the interval I_{k_j}. The solution of the IDE, which is denoted as $I_y(t)$, represents the set of all the ODE solutions obtained by considering all the possible combinations of the uncertain parameters and the initial conditions $I_y(0)$.

The most straightforward method for calculating the numerical solution of an IDE involves extending the numerical algorithms designed for solving a real-valued ODE with the use of IA-based operators, which allow computing ranges for numerical computation, assuming that each real quantity x is considered "unknown but bounded" within an interval of real numbers $X = [x_{inf}, x_{sup}]$. The deployment of IA-based operators allows estimating a reliable interval solution, called the outer solution, which is guaranteed to include the exact solution for each combination of the parameters uncertainty.

A more accurate estimation of the IDE solution set can be computed by using AA, which is a methodology developed in the context of reliable computing that allows modelling the errors propagation in long calculation chains. While retaining the simplicity of IA, AA can keep track of the statistical correlation between interval variables throughout the calculations.

2.7.2 IA-based consensus protocols

To model the effects of data uncertainty in achieving consensus, we analyze the application of IA-based computing. This approach is well suited for solving decentralized monitoring and estimation problems as it can represent the accuracy of computed quantities as an inherent part of the computation process, without necessitating information about the probabilistic structure of parameter uncertainties. The integration of IA in decentralized protocols requires modeling each state variable by an interval of floating-point numbers without a probability structure. These intervals are manipulated (added, subtracted, and/or multiplied) in a manner ensuring that each computed interval reliably includes the unknown value of the represented quantity. Following the introduction of fundamental aspects of IA-based computing, the formalization of IA-based decentralized consensus protocols is analyzed.

2.7.2.1 Elements of IA-based computing

IA is a range-based paradigm for reliable numerical analysis that allows describing each uncertain quantity x by an interval of real numbers $\bar{x}$, which is guaranteed to contain the (unknown) "true" value of x [23].

More specifically, a real interval is a compact set, which is defined as [24]

$$\bar{x} = \lfloor x_{lo}, x_{hi} \rfloor = \{x \in R : x_{lo} \leq x \leq x_{hi}\} \tag{2.45}$$

where x_{lo} and x_{hi} are the lower and upper bound of the interval, respectively.

These intervals are amalgamated and manipulated using proper set-based operators, which generalize real-value functions to handle interval-based uncertain variables, ensuring that the computed intervals include the sets of all the values of the calculated quantities. For this purpose, IA defines for each mathematical function $z = f(x_1, .., x_n)$ with $x_i \in R$ being the corresponding interval extension $\bar{z} = \bar{f}(\bar{x}_1, .., \bar{x}_n)$, which allows estimating an interval $\bar{z}$ including all the values $z = f(x)$ for $(x_1, .., x_n)$ varying independently over the given intervals $(\bar{x}_1, .., \bar{x}_n)$.

Deriving interval extensions for linear functions is straightforward, involving the definition of a closed-form expression for extreme values as arguments independently vary over specified intervals.

Conversely, establishing interval extensions for nonlinear functions poses a substantial challenge. Formulating analytic formulas for the "exact" extreme values of these functions is often exceptionally intricate. To uphold the essential invariant of range analysis, a conservative (yet computationally tractable) approximation of the "exact" function range is necessary. In such instances, the resulting computed intervals serve as outer estimations of the "exact" solution sets.

The interval extensions of key mathematical operators and elementary functions can be combined to compute extensions for complex functions. This process employs the same mathematical schemes as real numbers computing. Consequently, any numerical algorithm designed for real numbers can seamlessly transition to processing interval variables by substituting real number operators and functions with their corresponding interval extensions. This approach enables the evolution of numerical algorithms from real number computation to interval computation, identifying solutions that are no longer deterministic (valid only for fixed real input data) but described by intervals. According to the fundamental invariant of range analysis, these intervals are guaranteed to encompass the real values of solutions for all instances of uncertain input data within the specified intervals.

The interval extensions of basic mathematical operators can be defined as follows:

$$\begin{aligned}
-\bar{x} &= [-x_{hi}, -x_{lo}] \\
\bar{x}+\bar{y} &= [x_{lo}+y_{lo}, x_{hi}+y_{hi}] \\
\bar{x}+c &= [x_{lo}+c, x_{hi}+c] \\
\bar{x}-\bar{y} &= [x_{lo}-y_{hi}, x_{hi}-y_{lo}] \\
c\cdot\bar{x} &= [c\cdot x_{lo}, c\cdot x_{hi}] c>0 \\
c\cdot\bar{x} &= [c\cdot x_{hi}, c\cdot x_{lo}] c<0
\end{aligned} \tag{2.46}$$

The interval extension of the product operator, $\bar{x}*\bar{y}$, asks for deriving an analytic expression of the minimum and maximum values of the function xy for (x,y) varying over the range $[x_{lo}, x_{hi}]\text{x}[y_{lo}, y_{hi}]$. For this purpose, it should be noted that the function xy is linear in y (x) for each fixed x (y). Consequently, the maximum and minimum values of this function, named a and b, respectively, are located at the corner of the rectangle $[x_{lo}, x_{hi}]\text{x}[y_{lo}, y_{hi}]$, namely:

$$\begin{aligned}
a &= \min\{x_{lo}\cdot y_{lo}, x_{lo}\cdot y_{hi}, x_{hi}\cdot y_{lo}, x_{hi}\cdot y_{hi}\} \\
b &= \max\{x_{lo}\cdot y_{lo}, x_{lo}\cdot y_{hi}, x_{hi}\cdot y_{lo}, x_{hi}\cdot y_{hi}\}
\end{aligned} \tag{2.47}$$

The same approach can be deployed in the task of computing the interval extension of the division, $\bar{x}/\bar{y}$, which can be expressed as the product of $\bar{x}$ by $1/\bar{y}$. Anyway, since the reciprocal function $1/y$ is not defined for $y=0$, particular care should be devoted in computing the interval extension by properly checked the following conditions:

$$\begin{aligned}
&\text{if } y_{lo}<0 \text{ and } y_{hi}>0 \text{ then } a=-\infty, b=+\infty \\
&\text{if } y_{hi}=0 \text{ then } a=-\infty \\
&\text{if } y_{hi}\neq 0 \text{ then } a=\frac{1}{y_{hi}} \\
&\text{if } y_{lo}=0 \text{ then } b=+\infty \\
&\text{if } y_{lo}\neq 0 \text{ then } b=\frac{1}{y_{lo}}
\end{aligned} \tag{2.48}$$

Other fundamental IA-based operators are the midpoint $m(\bar{x})$ and the radius $r(\bar{x})$ of the interval $\bar{x}$, which are defined as

$$\begin{aligned}
m(\bar{x}) &= \frac{x_{lo}+x_{hi}}{2} \\
r(\bar{x}) &= \frac{x_{hi}-x_{lo}}{2}
\end{aligned} \tag{2.49}$$

Moreover, it could be also important defining the intersection of two intervals $\bar{x}$ and $\bar{y}$, which is defined as

$$\bar{x}\cap\bar{y} = [\max\{x_{lo}, y_{lo}\}, \min\{x_{hi}, y_{hi}\}] \tag{2.50}$$

and the convex hull of two intervals $\bar{x}$ and $\bar{y}$, which is the smallest interval $\bar{x}\cup\bar{y}$ containing $[\min\{x_{lo}, y_{lo}\}, \max\{x_{hi}, y_{hi}\}]$.

Finally, in order to achieve consensus on vector observations, it could be useful introducing the concept of interval matrices $\bar{\mathbf{A}} = (\bar{a}_{ij})$ with $i \in [1,m]$ and $j \in [1,n]$, and interval vectors $\bar{\mathbf{x}} = (\bar{x}_i)$, with $i \in [1,n]$, which are defined as follows [20]:

$[\mathbf{A}_{lo}, \mathbf{A}_{hi}] = \{\mathbf{B} \in R^{mxn} : \mathbf{A}_{lo} \leq \mathbf{B} \leq \mathbf{A}_{hi}\}$

$[\mathbf{x}_{lo}, \mathbf{x}_{hi}] = \{\mathbf{z} \in R^{n} : \mathbf{x}_{lo} \leq \mathbf{z} \leq \mathbf{x}_{hi}\}$

Starting from these definitions, it is possible to express the product $\bar{\mathbf{A}}\bar{\mathbf{x}}$ as follows:

$$\bar{\mathbf{A}}\bar{\mathbf{x}} = \sum_{j=1}^{n} \bar{a}_{ij}\bar{x}_j \subseteq \{\mathbf{A}\mathbf{x} : \mathbf{A} \in \bar{\mathbf{A}}, \mathbf{x} \in \bar{\mathbf{x}}\} \tag{2.51}$$

Hence, $\bar{\mathbf{A}}\bar{\mathbf{x}}$ is the interval vector containing the left set in (2.51).

An interval vector enclosing some set S as tight as possible is called the (interval) hull of S [24].

2.7.2.2 Achieving decentralized consensus by IA-based computing

In this section, we introduce an IA-based methodology for addressing consensus problems in the presence of data uncertainty [21]. Specifically, each variable in the decentralized protocol is represented by an interval assumed to be constant during the acquisition period. This interval models the primary sources of data uncertainty affecting the corresponding variable, including measurement uncertainties, measurement system repeatability, communication latencies, and rounding errors. The measurement uncertainties are further characterized as the sum of instrumental measurement uncertainty (dependent on the metrological characteristics of the sensing equipment) and measurement system repeatability (typically derived from a preliminary statistical analysis) [18]. Consequently, the variables sensed by the dynamic agent can be delineated by the following interval:

$$\bar{\omega}_i^k = [\omega_{i,lo}^k, \omega_{i,hi}^k] \; k \in [1,L] \tag{2.52}$$

The computation of the state of the ith oscillator, denoted as $\bar{\theta}_i^k$, is achievable for each dynamic agent through the solution of the decentralized consensus problem using IA-based computing. This necessitates substituting each numerical operator found in the decentralized protocols with the corresponding interval operator and replacing the non-linear coupling function with its interval extension.

Specifically, assuming each of the N dynamic agents is equipped with a single sensor and the interaction between agents is instantaneous, without any propagation delay, the network of dynamic agents can address the consensus problem by implementing the following IA-based decentralized protocol:

$$\dot{\bar{\theta}}_i(t) = \bar{\omega}_i + \frac{K}{c_i} \sum_{j=1}^{N} F^a(\bar{\theta}_j(t) - \bar{\theta}_i(t)) \tag{2.53}$$

which can be approximated by the following finite difference equation:

$$\bar{\theta}_i(t + \Delta t) = \bar{\theta}_i(t) + \bar{\omega}_i \Delta t + \frac{K\Delta t}{c_i} \sum_{j=1}^{N} F^a(\bar{\theta}_j(t) - \bar{\theta}_i(t)) \tag{2.54}$$

Hence, when the sensors network reaches a consensus, all the sensors synchronize to the following interval:

$$\bar{\omega}^* = \frac{\sum_{i=1}^{N} c_i \bar{\omega}_i}{\sum_{i=1}^{N} c_i} \tag{2.55}$$

This IA-based consensus protocol enables the information shared among dynamic agents to inherently capture the propagation of data uncertainties as the network of agents achieves consensus. This attribute proves highly valuable in facilitating the prompt development of sensitivity analysis. It is noteworthy that the IA-based consensus protocol eliminates the need for assumptions regarding the maximum interval width of dynamic agent observations. Consequently, it can be effectively employed for the development of large-scale sensitivity analysis.

2.7.2.3 Numerical example

This section describes the implementation of the IA-based reliable protocols for decentralized computing within a 20 kV, 55-bus smart microgrid (SMG), incorporating interval uncertainties [21]. A self-organizing network, comprising 55 dynamic agents (one for each bus), has been deployed across the SMG. The coupling coefficients of the communication network topology among agents are defined based on the same topology as the electrical network, employing a linear coupling strategy between the built-in oscillators. For simplicity, we assume that all dynamic agents are equipped with a single sensor, and instantaneous interaction occurs between agents.

To evaluate the performance of this computing architecture in tackling complex tasks, the dynamic agents are tasked with solving the power flow problem. This problem involves calculating the steady-state SMG operating point based on a given set of measured data, such as load demand and real power generation. It serves as a crucial tool for various SMG control and monitoring functions related to electrical grid optimization, state estimation, and service restoration. The self-organizing agents network employs a decentralized solution algorithm based on the IA-based average consensus protocol. In this study, a $\pm 5\%$ tolerance for both measurement uncertainties and measurement system repeatability is assumed [18].

The trajectory bounds for the state of each dynamic agent are presented in Figures 2.9–2.12, which depicts the mean value and the tolerance of the real and the imaginary component of the bus voltage magnitude, and in Figures 2.13 and 2.14, which reports the mean value and the tolerance of the line currents magnitude, respectively.

The analysis of Figures 2.9–2.12 confirms that the IA-based protocol enables dynamic agents to solve the uncertain power flow problem in about 100 iterations. This capability allows the dynamic agents network to reach a consensus on key variables describing the actual power system operation state, identifying potential grid anomalies, and detecting incipient constraint violations that could compromise correct system operation.

Moreover, Figures 2.13 and 2.14 illustrate the agents state in computing the current magnitudes on all power lines. Leveraging this information, each dynamic

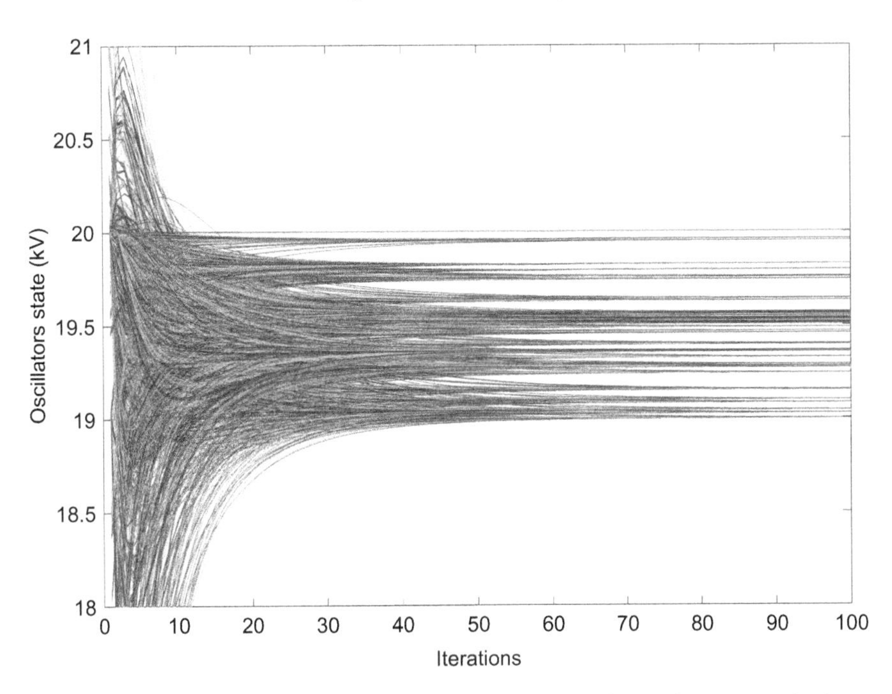

Figure 2.9 Mean value of the real component of the bus voltage magnitudes

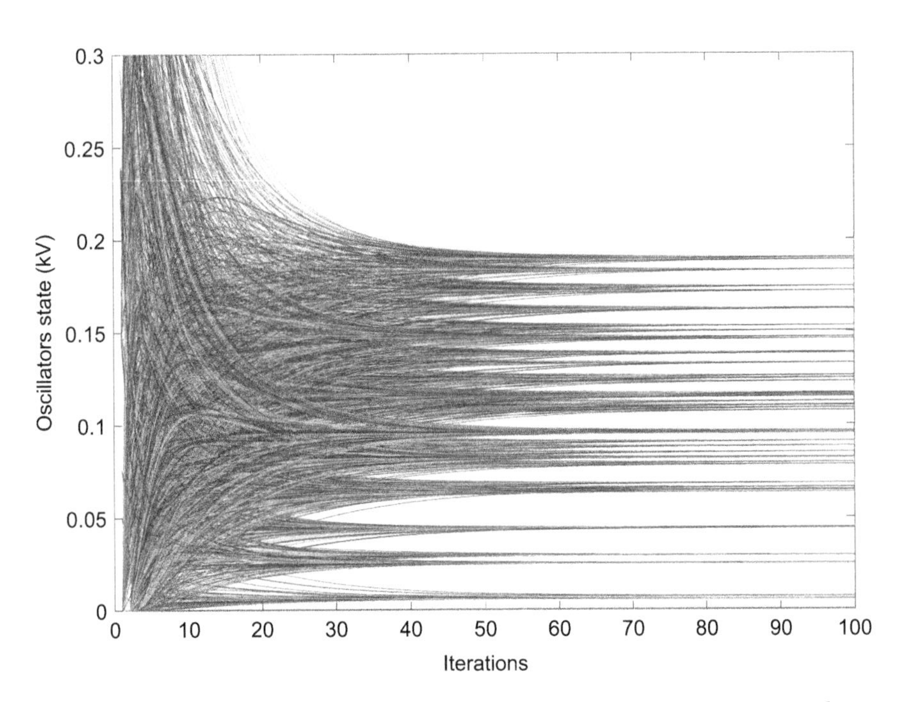

Figure 2.10 Tolerance of the real component of the bus voltage magnitudes

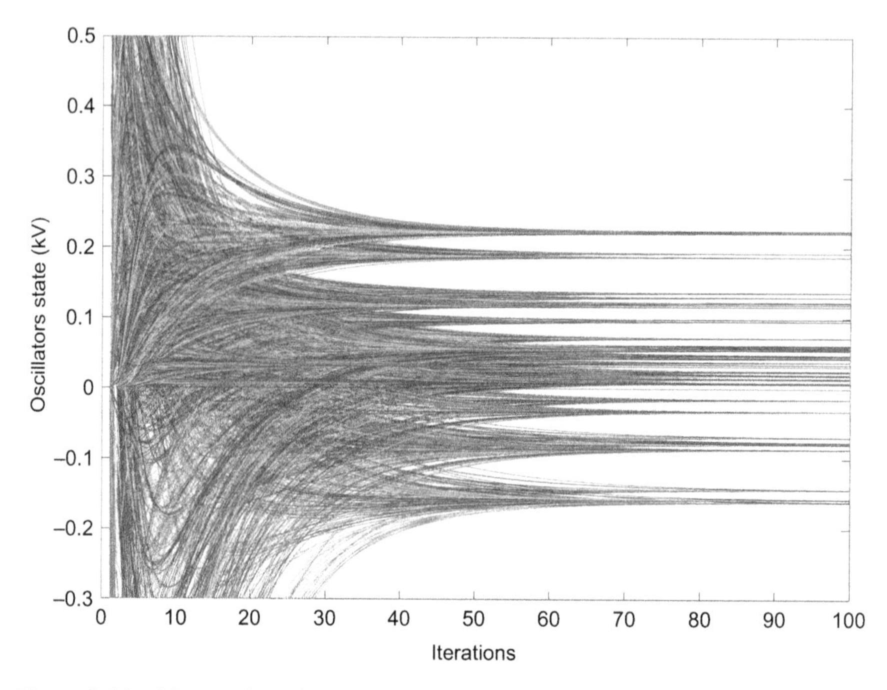

Figure 2.11 Mean value of the imaginary component of the bus voltage magnitudes

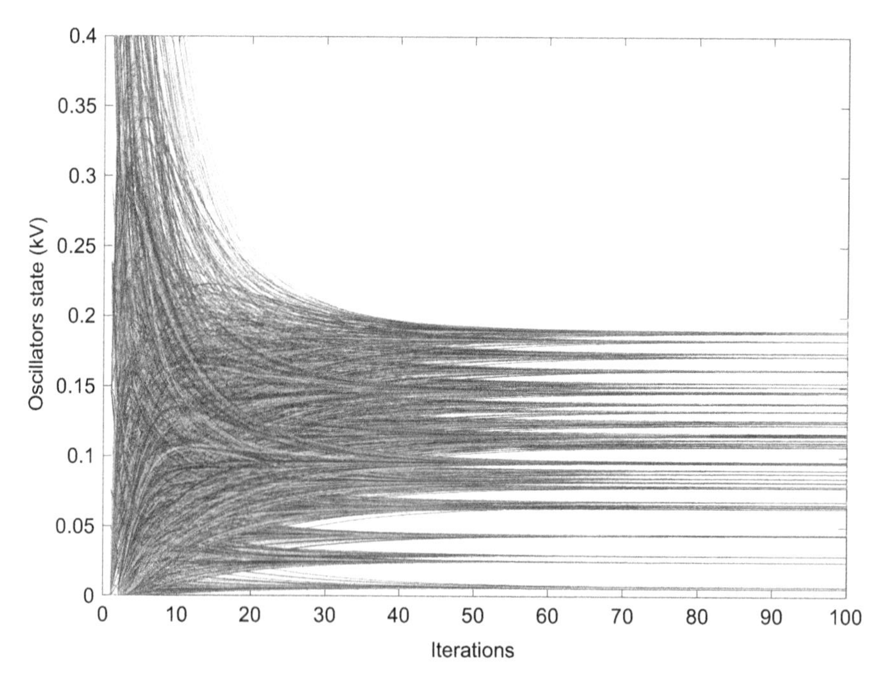

Figure 2.12 Tolerance of the imaginary component of the bus voltage magnitude

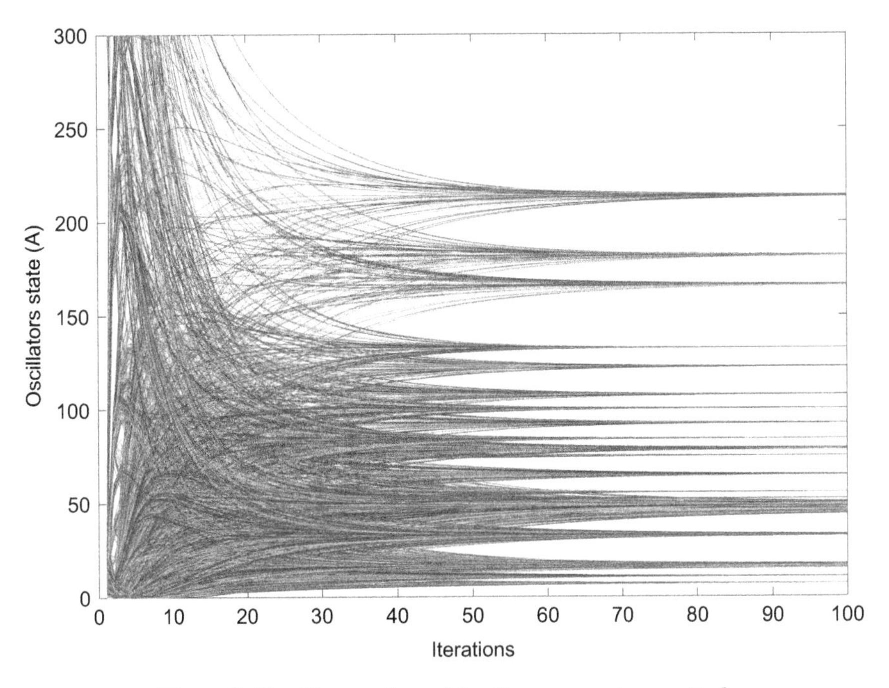

Figure 2.13 Mean value of the lines current magnitude

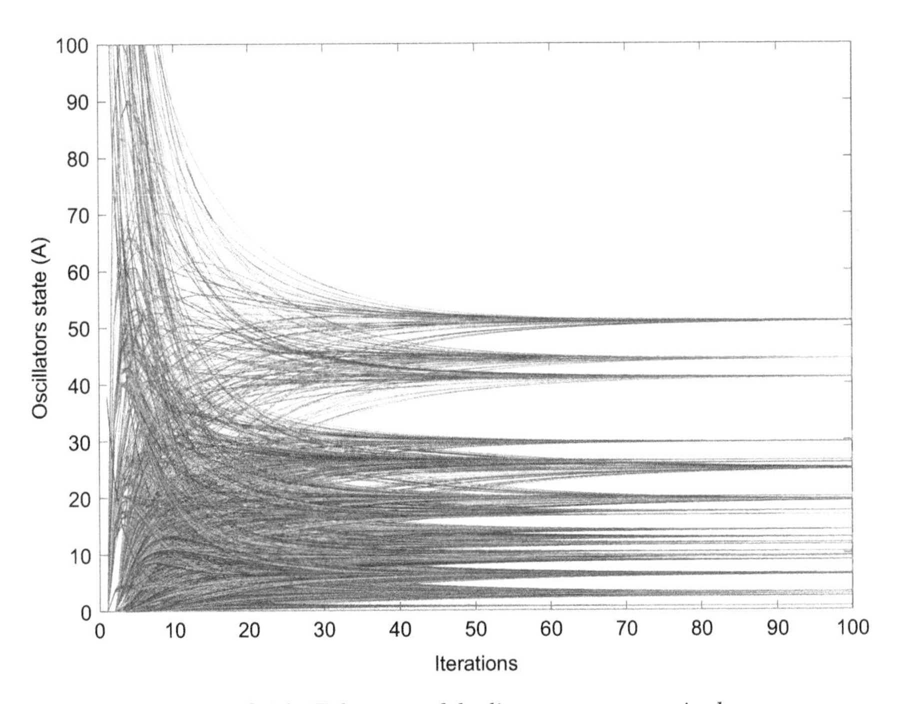

Figure 2.14 Tolerance of the lines current magnitude

agent can compute the available thermal capability of the entire power network, considering the impact of all data uncertainty affecting grid operation.

To assess the effectiveness of the power flow solution computed by the dynamic agents network, the obtained solution bounds are compared to those obtained using a centralized solution approach based on the Monte Carlo method. For this comparison, 2000 different samples of input variables within the assumed input bounds were randomly selected, and a conventional solution algorithm was applied to compute the corresponding power flow solutions. It is important to note that this comparative analysis does not aim to demonstrate superior performance of IA-based consensus protocols compared with a centralized Monte Carlo solution algorithm, as the latter is typically assumed to yield correct solution intervals. The results of the comparative analysis are summarized in Figures 2.15 and 2.16.

Analyzing this data reveals that the solution bounds computed by the IA-based consensus protocol are slightly conservative compared to the solution obtained by applying the Monte Carlo simulation. However, the latter requires a detailed model of the electrical grid and a central fusion center for acquiring and processing all sensor grid measurements. In contrast, the dynamic agents network solves the power flow problem using a fully decentralized/non-hierarchical computing paradigm. The problem solution is assessed by the agents network by processing local variables only, without the need for a data fusion center. This makes this approach distributed, self-organized, and highly scalable.

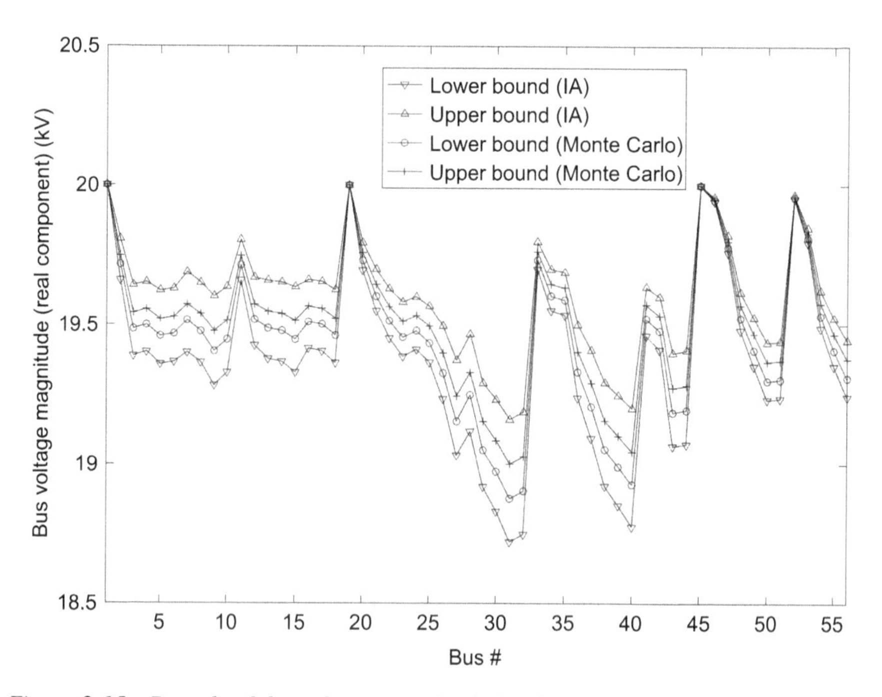

Figure 2.15 Bounds of the voltage magnitude (real component) computed by the dynamic agents and by a Monte Carlo simulation (2000 trials)

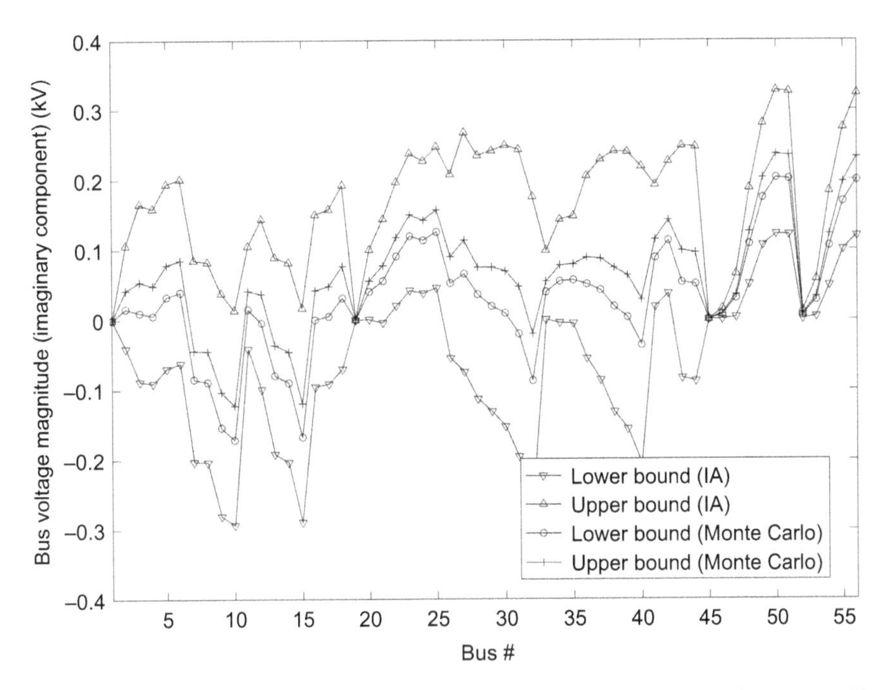

Figure 2.16 Bounds of the voltage magnitude (imaginary component) computed by the dynamic agents and by a Monte Carlo simulation (2000 trials)

2.7.3 AA-based consensus protocols

While the inclusion property of IA-based computing allows reliably solving consensus problems in the presence of multiple and correlated uncertainties, the obtained outer solutions may not always provide the expected level of informativeness. A study in [22] reveals that employing IM to solve interval differential equations can lead to unconventional solutions. This occurs because the IM formalism lacks the capability to represent the interaction established by the differential equation between variables. The inability to depict this interaction permits the introduction of spurious values into the solution, potentially causing the system's evolution to reach regions where a correct solution does not exist—referred to as the "wrapping problem." This phenomenon is familiar in the simulation of qualitative systems [25] and necessitates specific solutions. In [22], these challenges are addressed by defining reliable consensus protocols based on the use of AA.

AA is a computing paradigm for range analysis, which can handle both external sources of error (e.g., imprecise or missing input data, uncertainty in mathematical modeling) and internal sources of error (e.g., roundoff and truncation errors) [25,26]. While similar to IA, AA goes further by capturing any correlation between inputs and computed quantities. This capability allows AA to determine an output within a narrower interval, thereby reducing the likelihood of encountering the error explosion problems observed in extensive IA computations.

2.7.3.1 Elements of AA-based computing

AA is a methodology devised within the framework of interval analysis, focusing on managing the propagation of errors in extensive calculation sequences. Despite maintaining the simplicity of interval arithmetic, AA has the capability to model the correlation between interval variables throughout the computational process.

An interval variable $\hat{x}$, characterized by uncertainty linked to n independent sources of uncertainty, can be represented as a first-order polynomial known as an affine form [25,26]:

$$\hat{x} = x_0 + x_1\varepsilon_1 + x_2\varepsilon_2 + \cdots + x_n\varepsilon_n + x_e\varepsilon_e \tag{2.56}$$

An affine form is defined by a central value, denoted as x_0, which remains unaffected by uncertainty, and a set of partial deviations $x_i\ \forall i \in [1,...,n]$. Each partial deviation is multiplied by a corresponding noise symbol ε_i, representing an independent source of uncertainty. The noise symbols ε_i are random variables uniformly distributed in the range $[-1,1]$. The last term, $x_e\varepsilon_e$, represents an approximation error term known as endogenous uncertainty, which model nonlinear dependencies.

AA serves as a generalization of interval arithmetic. Specifically, for a given interval variable $\hat{p} = [\underline{p},\overline{p}]$, the corresponding affine form $\hat{q}$ can be computed as follows:

$$q_0 = \frac{\underline{p}+\overline{p}}{2} \quad q_{new} = \frac{\overline{p}-\underline{p}}{2} \tag{2.57}$$

Hence, if $\hat{p} = [2,8]$, the corresponding affine form would be $\hat{q} = 5 + 3\varepsilon_1$.

The basic AA-based operations, whose outputs are exact affine representations, include addition, subtraction, and scalar multiplication. Given two affine forms $\hat{x}$ and $\hat{y}$, these operations are

$$\hat{x} \pm \hat{y} = (x_0 \pm y_0) + \sum_{i=1}^{n}(x_i \pm y_i)\varepsilon_i + (x_e + y_e)\varepsilon_e \tag{2.58}$$

$$\alpha\hat{x} = \alpha x_0 + \sum_{i=1}^{n}\alpha x_i\varepsilon_i + \alpha x_e\varepsilon_e \tag{2.59}$$

Nonlinear operations on affine forms, like multiplication or division, do not yield an affine combination of the “primitive” noise symbols. Therefore, these functions are approximated using an affine form, and the error term of the approximation depends on the chosen method of approximation. In particular, given a nonlinear function of affine forms $f^*(\varepsilon_1,\varepsilon_2,...,\varepsilon_n)$, an affine approximation $f^\alpha(\varepsilon_1,\varepsilon_2,...,\varepsilon_n)$ would be [25]

$$f^\alpha(\varepsilon_1,\varepsilon_2,...,\varepsilon_n) = z_0 + z_1\varepsilon_1 + z_2\varepsilon_2 + \ldots + z_n\varepsilon_n + z_e\varepsilon_e \tag{2.60}$$

The term $z_e\varepsilon_e$ is the error term, and z_e represents the difference between the real value of the function and its approximation:

$$z_e = f^\alpha(\varepsilon_1,\varepsilon_2,...,\varepsilon_n) - f^*(\varepsilon_1,\varepsilon_2,...,\varepsilon_n) \tag{2.61}$$

Enhancing the robustness of the method requires an error term ensuring that $f^\alpha(\varepsilon_1,\varepsilon_2,...,\varepsilon_n)$ serves as an upper bound for the actual function.

In instances of iterative operations, error terms may exert greater influence as each operation contributes to their increase. However, under specific conditions of the problem and with the appropriate approximation of the non-linear function, it is conceivable that the error term could converge to a finite value.

In solving decentralized consensus problems could be useful introducing affine vectors and matrices, which can be defined by building upon the definition of scalar affine forms.

An $(M \times P)$ affine matrix, represented as $\hat{A}$, involving M uncertainty sources, assuming that all nonlinearities are merged into a single error term, can be expressed as

$$\hat{A} = A_0 + A_1\varepsilon_1 + \cdots + A_n\varepsilon_n + A_e\varepsilon_e \tag{2.62}$$

A single element $\hat{a}_{ij}$ of the matrix $\hat{A}$ would be

$$\hat{a}_{ij} = a_{i,j}^{(0)} + a_{i,j}^{(1)}\varepsilon_1 + \cdots + a_{i,j}^{(n)}\varepsilon_n + a_{i,j}^{(e)}\varepsilon_e \tag{2.63}$$

The explicit form of the matrix is the following:

$$\begin{aligned}\hat{A} = &\begin{bmatrix} a_{1,1}^{(0)} & a_{1,2}^{(0)} & \dots & a_{1,P}^{(0)} \\ a_{2,1}^{(0)} & a_{2,2}^{(0)} & \dots & a_{2,P}^{(0)} \\ & \dots & \dots & \\ a_{M,1}^{(0)} & a_{M,2}^{(0)} & \dots & a_{M,P}^{(0)} \end{bmatrix} + \\ &+ \begin{bmatrix} a_{1,1}^{(1)} & a_{1,2}^{(1)} & \dots & a_{1,P}^{(1)} \\ a_{2,1}^{(1)} & a_{2,2}^{(1)} & \dots & a_{2,P}^{(1)} \\ & \dots & \dots & \\ a_{M,1}^{(1)} & a_{M,2}^{(1)} & \dots & a_{M,P}^{(1)} \end{bmatrix} \varepsilon_1 + \dots + \\ &+ \begin{bmatrix} a_{1,1}^{(n)} & a_{1,2}^{(n)} & \dots & a_{1,P}^{(n)} \\ a_{2,1}^{(n)} & a_{2,2}^{(n)} & \dots & a_{2,P}^{(n)} \\ & \dots & \dots & \\ a_{M,1}^{(n)} & a_{M,2}^{(n)} & \dots & a_{M,P}^{(n)} \end{bmatrix} \varepsilon_n = \bar{A}_0 + \sum_{k=1}^{n} \bar{A}_k\varepsilon_k \end{aligned} \tag{2.64}$$

Given the vector $\varepsilon = \begin{bmatrix} 1 & \varepsilon_1 & \dots & \varepsilon_n \end{bmatrix}$ and the matrix $\mathrm{vec}(A) = \begin{bmatrix} A_0 & A_1 & \dots & A_n \end{bmatrix}$, the affine matrix $\hat{A}$ can be expressed as a $\langle \mathrm{vec}(A), \varepsilon \rangle_F$ [27].

For example, given the affine matrix $\hat{B}$

$$\hat{B} = \begin{bmatrix} 1 + 2\varepsilon_1 + 4\varepsilon_2 + 6\varepsilon_3 & 2 + 8\varepsilon_2 \\ 3 + 4\varepsilon_3 & 9\varepsilon_1 + 7\varepsilon_3 \end{bmatrix} \tag{2.65}$$

The coefficient matrices for each noise symbol would be

$$\begin{aligned} \hat{B}_0 &= \begin{bmatrix} 1 & 2 \\ 3 & 0 \end{bmatrix} \; \hat{B}_1 = \begin{bmatrix} 2 & 0 \\ 0 & 9 \end{bmatrix} \\ \hat{B}_2 &= \begin{bmatrix} 4 & 8 \\ 0 & 0 \end{bmatrix} \; \hat{B}_3 = \begin{bmatrix} 6 & 0 \\ 4 & 7 \end{bmatrix} \end{aligned} \tag{2.66}$$

The affine matrix $\hat{B}$ can also be defined as $\langle \mathrm{vec}(B), \varepsilon \rangle_F$.

2.7.3.2 Achieving decentralized consensus by AA-based computing

AA-based computing can be adopted for modeling data uncertainty in solving decentralized consensus problems [22]. With AA, each parameter of the decentralized protocols is defined by a central value and a set of partial deviations. According to this approach, the statistical features of the measured quantities are assumed to remain constant during the acquisition period. Specifically, the measurement uncertainties are modeled as the sum of two primary components: the first being a priori knowledge of the instrumental measurement uncertainty (dependent on the metrological characteristics of the sensing equipment), and the second being the measurement system repeatability (typically derived through statistical analysis of measured quantities under defined measurement conditions) [18].

Consequently, the observations sensed by the ith agent can be expressed using the following affine form:

$$\hat{\omega}_i^k = \omega_{i,0}^k + \omega_{i,A}^k \varepsilon_A + \omega_{i,B}^k \varepsilon_B + \omega_{i,C}^k \varepsilon_C + \omega_{i,D}^k \varepsilon_D \; k \in [1,L] \tag{2.67}$$

where

- $\varepsilon_A, \varepsilon_B, \varepsilon_C, \varepsilon_D$ are noise symbols describing the measurement uncertainties, the measurement system repeatability, the communication latencies, and the rounding errors, respectively;
- $\omega_{i,A}^k$, $\omega_{i,B}^k$, $\omega_{i,C}^k$, $\omega_{i,D}^k$ corresponding partial deviations that characterize the kth sensor of the ith agent;
- $\omega_{i,0}^k$ is the central value of the variable sensed by the kth sensor of the ith agent.

Hence, the corresponding state of the ith oscillator can be computed as follows:

$$\hat{x}_i^k(t) = x_{i,0}^k(t) + x_{i,A}^k(t)\varepsilon_A + x_{i,B}^k(t)\varepsilon_B + x_{i,C}^k(t)\varepsilon_C + x_{i,D}^k(t)\varepsilon_D + x_{i,NA}^k(t)\varepsilon_{NA} \; k \in [1,L] \tag{2.68}$$

where the new noise symbol ε_{NA} models the effect of the approximation errors.

The central values $x_{i,0}^k(t)$ and the partial deviations ($x_{i,A}^k(t), x_{i,B}^k(t), x_{i,C}^k(t), x_{i,D}^k(t)$, $x_{i,NA}^k(t)$) for $k \in [1,L]$ and $i \in [1,N]$ can be computed individually by each agent using an AA-based representation to formalize the decentralized consensus protocols. This involves substituting each numerical operator in the protocols with the corresponding AA operator and approximating the nonlinear coupling function with an appropriate affine representation. This robust approach ensures that the information exchanged among agents inherently encompasses the impact of data uncertainties when the decentralized network of agents achieves consensus. It is essential to note that according to this protocol, all the agents reach a consensus by synchronizing to an affine form rather than a deterministic value.

In particular, if we consider each of the N agents to have a single sensor (i.e., $L = 1$) and assume instantaneous interaction between agents without propagation delay, we can express the observation and state of the ith oscillator through the following affine forms:

$$\begin{aligned} \hat{\omega}_i &= \omega_{i,0} + \omega_{i,A}\varepsilon_A + \omega_{i,B}\varepsilon_B + \omega_{i,C}\varepsilon_C + \omega_{i,D}\varepsilon_D \\ \hat{x}_i(t) &= x_{i,0}(t) + x_{i,A}(t)\varepsilon_A + x_{i,B}(t)\varepsilon_B + x_{i,C}(t)\varepsilon_C + x_{i,D}(t)\varepsilon_D + x_{i,NA}(t)\varepsilon_{NA} \end{aligned} \tag{2.69}$$

Hence, the agents network can solve the average consensus problem by deploying the following AA-based protocol:

$$\dot{\hat{x}}_i(t) = \hat{\omega}_i + \frac{K}{q_i}\sum_{j=1}^{N} f^a(\hat{x}_j(t) - \hat{x}_i(t)) \tag{2.70}$$

which can be approximated by the following finite difference equation:

$$\hat{x}_i(t+\Delta t) = \hat{x}_i(t) + \hat{\omega}_i\Delta t + \Delta t\frac{K}{q_i}\sum_{j=1}^{N} f^a(\hat{x}_j(t) - \hat{x}_i(t)) \tag{2.71}$$

In this case, if the oscillators reach a consensus, they synchronize to the following affine form:

$$\hat{\omega}^* = \frac{\sum_{i=1}^{N} c_i\hat{\omega}_i}{\sum_{i=1}^{N} c_i} + \omega_{NA}(t)\varepsilon_{NA} \tag{2.72}$$

Here, the expression $\omega_{NA}(t)\varepsilon_{NA}$ encapsulates the affine approximation errors arising from the affine function $f^a(\hat{x})$.

This scheme enables each agent to precisely determine the minimum interval that encompasses all potential values of the synchronization state when factoring in the set of uncertainties, namely

$$[\omega^*] = [\omega_0^* - \lambda, \omega_0^* + \lambda] \tag{2.73}$$

where λ is the radius of the affine forms defined as

$$\lambda = |\omega_A^*| + |\omega_B^*| + |\omega_C^*| + |\omega_D^*| + |\omega_{NA}^*| \tag{2.74}$$

This expression also facilitates the identification of the most relevant uncertainties impacting the synchronization state. Additionally, it provides a means to explicitly discern the relative impact of each specific source of uncertainty. Such insights prove invaluable for performing sensitivity analysis. It is noteworthy that this AA-based protocol does not necessitate presumptions about the maximum width of partial deviations in agent observations. Therefore, it can be applied effectively in the development of extensive sensitivity analyses.

2.7.3.3 Numerical example

This section describes the application of the described AA-based consensus protocol in the task of monitoring the IEEE 118-bus test system, considering uncertainties [22]. A network of 118 collaborative dynamic agents has been deployed throughout the power network, with each agent assigned to a specific bus. The coupling coefficients governing the topology of the agent communication network are determined based on the electrical network's topology. For simplicity, it is assumed that all agents have a single sensor, and interactions occur instantaneously without propagation delay. The sensor is characterized by an overall accuracy of ±(0.04% of reading + 0.025% full scale) with a full scale of 690 V. Synchronization bounds obtained through the AA-based consensus protocol are compared with those derived from a Monte Carlo simulation, often considered to produce the "correct" solution intervals. The Monte Carlo simulation involves randomly selecting 1000 values within the assumed input bounds, applying a conventional consensus protocol to each, and determining interval solutions defined by the largest and smallest values of monitored variables.

A vector-based approach is employed for the AA representation of oscillators and consensus protocols. A ±5% tolerance for both measurement uncertainties and measurement system repeatability is assumed, ensuring a sufficiently wide interval for proper method evaluation.

Utilizing these data uncertainties, the AA-based average consensus protocol is applied in the task of estimating bounds for average active power demand, assuming a linear coupling strategy. Trajectory bounds for each agent state are illustrated in Figure 2.17, indicating convergence to two synchronization states representing lower and upper bounds of active power demand. Comparison with Monte Carlo results is depicted in Figures 2.18 and 2.19, with Figure 2.18 presenting the envelope of agent state trajectories and Figure 2.19 illustrating trajectory bounds for a specific bus. The AA-based linear consensus protocols demonstrate excellent approximation of trajectory bounds when compared to benchmark intervals obtained using the Monte Carlo approach. This is attributed to AA's intrinsic characteristics, allowing it to include correlations between agent states without generating spurious trajectories. Moreover, in this case, there are no affine approximation errors as all elementary operations required for solving the consensus problem can be precisely computed by an equivalent affine operation.

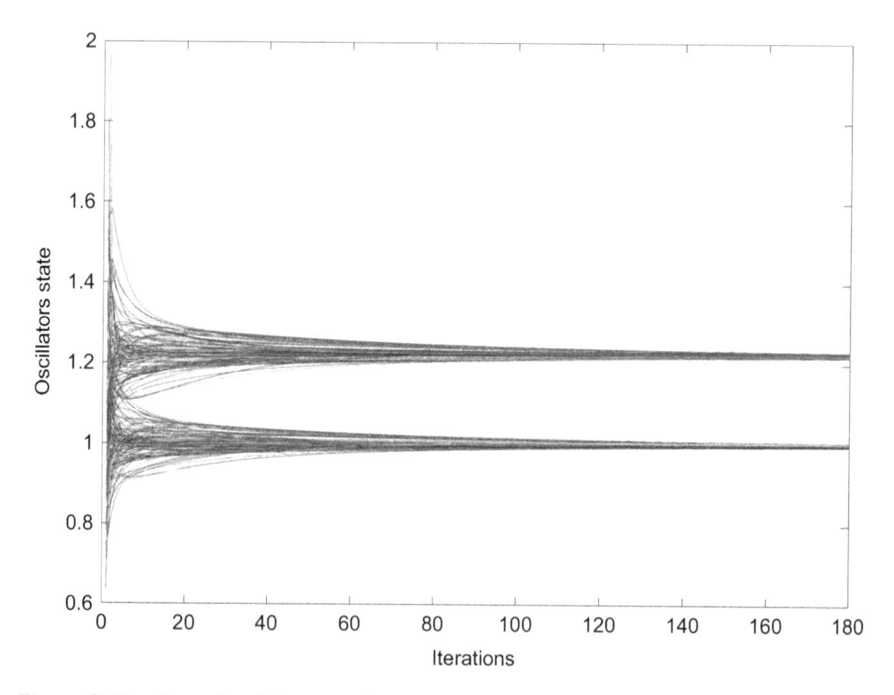

Figure 2.17 Bounds of the state derivatives computed by the AA-based linear consensus protocol

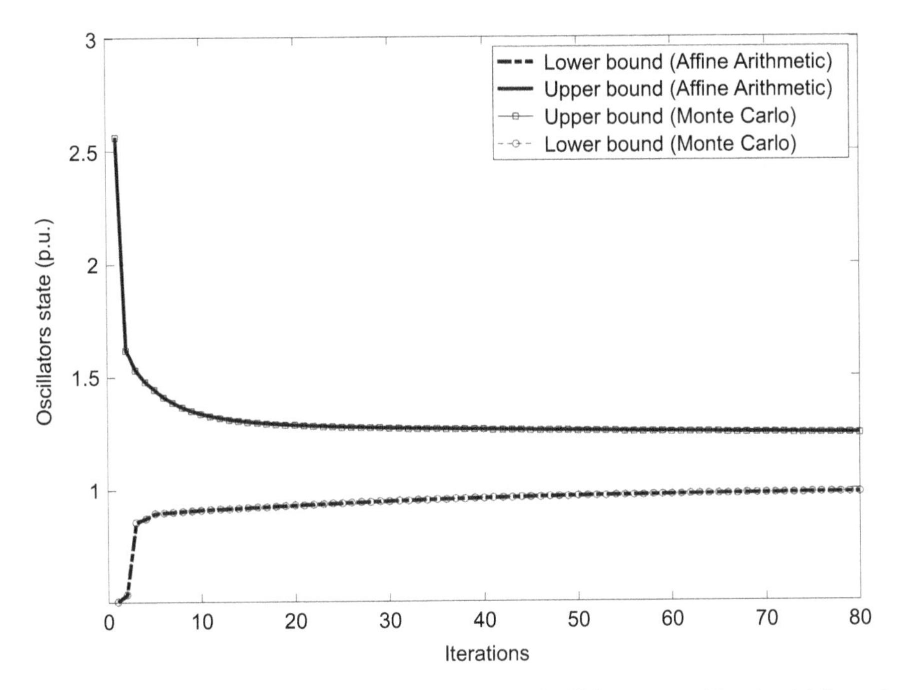

Figure 2.18 Trajectory bounds for the sensor node #25 computed by the AA-based linear consensus protocol and by a 1000-trial-based Monte Carlo model

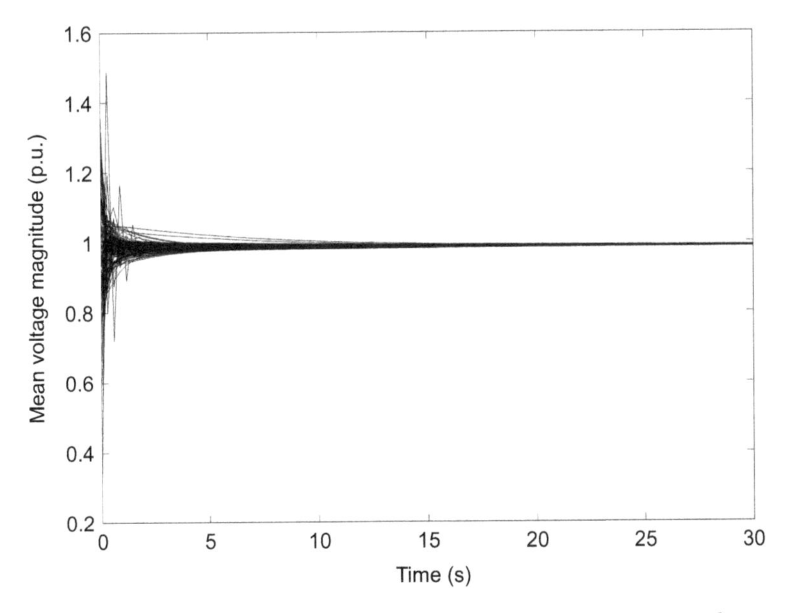

Figure 2.19 Evolution of the state derivatives in the task of computing the average grid voltage magnitude

References

[1] Liu J, Zhao Z, Ji J, *et al.* Research and application of wireless sensor network technology in power transmission and distribution system. *Intelligent and Converged Networks*. 2020;1(2):199–220.

[2] Bagherzadeh L, Shahinzadeh H, Shayeghi H, *et al.* Integration of Cloud Computing and IoT (CloudIoT) in smart grids: benefits, challenges, and solutions. In: *2020 International Conference on Computational Intelligence for Smart Power System and Sustainable Energy (CISPSSE)*; 2020. p. 1–8.

[3] Giridhar A and Kumar PR. Toward a theory of in-network computation in wireless sensor networks. *IEEE Communications Magazine*. 2006;44(4): 98–107.

[4] Barbarossa S and Scutari G. Decentralized maximum-likelihood estimation for sensor networks composed of nonlinearly coupled dynamical systems. *IEEE Transactions on Signal Processing*. 2007;55(7):3456–3470.

[5] Khan S. Distributed sensors, computation and AI for automation, protection and maintenance of power grid. In: *2022 18th International Computer Engineering Conference (ICENCO)*. vol. 1; 2022. p. 130–135.

[6] di Bisceglie M, Galdi C, Vaccaro A, *et al.* Cooperative sensor networks for voltage quality monitoring in smart grids. In: *2009 IEEE Bucharest PowerTech*; 2009. p. 1–6.

[7] Vaccaro A, Velotto G, and Zobaa AF. A decentralized and cooperative architecture for optimal voltage regulation in smart grids. *IEEE Transactions on Industrial Electronics*. 2011;58(10):4593–4602.

[8] Iacoviello A, Loia V, Pietrosanto A, *et al.* Decentralized consensus protocols: the enabler for smarter grids monitoring. In: *2013 27th International Conference on Advanced Information Networking and Applications Workshops*; 2013. p. 1559–1564.

[9] Loia V and Vaccaro A. Decentralized economic dispatch in smart grids by self-organizing dynamic agents. *IEEE Transactions on Systems, Man, and Cybernetics: Systems*. 2014;44(4):397–408.

[10] Qamar H, Qamar H, and Vaccaro A. Design of fuzzy logic controllers for decentralized voltage regulation in grid connected photovoltaic systems. In: *2017 IEEE International Conference on Fuzzy Systems (FUZZ-IEEE)*; 2017. p. 1–6.

[11] Carlini EM, Giannuzzi GM, Pisani C, *et al.* Experimental deployment of a self-organizing sensors network for dynamic thermal rating assessment of overhead lines. *Electric Power Systems Research*. 2018;157:59–69.

[12] Pescosolido L, Barbarossa S, and Scutari G. Decentralized detection and localization through sensor networks designed as a population of self-synchronizing oscillators. In: *2006 IEEE International Conference on Acoustics Speech and Signal Processing Proceedings*. vol. 4; 2006. p. IV–IV.

[13] Hoppensteadt FC and Izhikevich EM. Pattern recognition via synchronization in phase-locked loop neural networks. *IEEE Transactions on Neural Networks*. 2000;11(3):734–738.

[14] Kay SM. *Fundamentals of Statistical Signal Processing, Vol I: Estimation Theory*. 1st ed. Upper Saddle River, NJ: Prentice Hall; 1993.

[15] Godsil C and Royle G. *Algebraic Graph Theory*. 1st edn. New York, NY: Springer; 2001.

[16] Wu CW. Synchronizability of networks of chaotic systems coupled via a graph with a prescribed degree sequence. *Physics Letters A*. 2005;346(4): 281–287.

[17] Wang XF and Chen G. Synchronization in scale-free dynamical networks: robustness and fragility. *IEEE Transactions on Circuits and Systems I: Fundamental Theory and Applications*. 2002;49(1):54–62.

[18] Capriglione D, Ferrigno L, Paciello V, *et al*. On the performance of consensus protocols for decentralized smart grid metering in presence of measurement uncertainty. In: *2013 IEEE International Instrumentation and Measurement Technology Conference (I2MTC)*; 2013. p. 1176–1181.

[19] Andreotti A, Caiazzo B, Petrillo A, *et al*. Robust finite-time voltage restoration in inverter-based microgrids via distributed cooperative control in presence of communication time-varying delays. In: *2020 IEEE International Conference on Environment and Electrical Engineering and 2020 IEEE Industrial and Commercial Power Systems Europe (EEEIC/I&CPS Europe)*; 2020. p. 1–6.

[20] Scutari G, Barbarossa S, and Pescosolido L. Distributed decision through self-synchronizing sensor networks in the presence of propagation delays and asymmetric channels. *IEEE Transactions on Signal Processing*. 2008;56(4):1667–1684.

[21] Formato G, Troiano L, and Vaccaro A. Achieving consensus in self-organizing multi agent systems for smart microgrids computing in the presence of interval uncertainty. *Journal of Ambient Intelligence and Humanized Computing*. 2014;5:821–828.

[22] Loia V, Terzija V, Vaccaro A, *et al*. An affine-arithmetic-based consensus protocol for smart-grid computing in the presence of data uncertainties. *IEEE Transactions on Industrial Electronics*. 2015;62(5):2973–2982.

[23] Moore RE. Interval Analysis. New Jersey, Englewood Cliffs, Prentice-Hall. 1966.

[24] Vaccaro A. *Interval Methods for Uncertain Power System Analysis*. New York, NY: Wiley, IEEE Press; 2023.

[25] Stolfi J and de Figueiredo LH. An introduction to affine arithmetic. *TEMA-Tendências em Matemática Aplicada e Computacional*. 2003;4(3):297–312.

[26] Stolfi J and De Figueiredo LH. Self-validated numerical methods and applications. In: *Proceedings of the Monograph for 21st Brazilian Mathematics Colloquium*. Citeseer; 1997.

[27] Pepiciello A, Fabrizio De Caro, Vaccaro A, and Djokic S. Affine Arithmetic-Based Reliable Estimation of Transition State Boundaries for Uncertain Markov Chains. *Electric Power Systems Research*. 2022;204:107711.

Chapter 3
Self-organizing wide-area measurement systems

Conventional electrical power systems were operated based on a "static model," wherein large and programmable power generation plants produced electricity, which was then transmitted to consumers via long transmission lines and wide distribution systems. In this context, the focus of the grid technological investments was mainly oriented on the task of enhancing the transmission system, which was considered the most strategic asset allowing the secure and reliable power system operation. This emphasis stemmed largely from the concentration of power generators in cost-effective locations. More specifically, substantial resources were allocated to constructing new transmission lines, with the goal of augmenting network reliability by increasing the grid meshing. This approach aimed at enabling the efficient transfer of large power transactions over large distances, while minimizing losses. Meanwhile, distribution networks received relatively less attention within the electricity supply chain, and they substantially played only a passive role. Consequently, the principles and criteria governing the planning and operation of distribution systems remained mainly unaltered over the overall power system history.

Currently, modern electricity grids are radically changing under the pressure of the new environmental policies, which aim at decarbonizing the overall power sector by replacing the large conventional power generators with distributed and dispersed renewable generating units. In this emerging scenario, many complex issues should be addressed such as the necessity of enhancing the grid flexibility in order to increase the hosting capacity of renewable generation facilities, enhancing distribution grid to reliably support bi-directional power flows, coordinating a large number of small distributed energy resources, and managing the distribution grid congestion by properly balancing grid ownership versus system operation and reliability coordination. These challenging issues are multi-faceted, and entail effectively balancing between ensuring resource adequacy, enhancing grid reliability, optimizing economic operation, and fulfilling strict environmental objectives. Moreover, assuring the reliable operation of future electrical grids is becoming more demanding due to the persistent evolution of power system functions, transitioning from operational jurisdiction to control responsibilities. This transformation is compounded by the increasing demand for reliability. To tackle these intricate challenges, power system operators should squarely confront the subsequent pivotal concerns:

- the rising number of network interconnections, which contributes to the heightened vulnerability of power networks to dynamic perturbations. This effect

is exacerbated as an increasing number of power components approach or reach their loading limits, resulting in more frequent occurrences of emergency operation states;
- the proliferation of smaller, widely dispersed generators, which has the potential to significantly augment the volume and the number of power transactions;
- the need for accurately predicting and modeling the behavior of electricity market operators, driven predominantly by unpredictable economic dynamics, which introduces substantial uncertainty into short-term power system operation;
- the growing necessity of performing comprehensive security analyses, which allows assessing the impact of multiple contingencies on the power network (e.g., N-2 security criteria);
- the massive integration of renewable energy-powered generating units, which introduces various side effects on power system operation, such as highly variable power injections and perturbations in bus voltage magnitude profiles.

In an effort to tackle these challenges, the smart grids concept has gained recognition as a highly promising enabling technology. The extensive implementation of this emerging framework is expected to augment the efficiency of existing electrical grids and the utilization of cleaner energy sources. This enhancement stems from the integration of computational intelligence methods, data analytic-based tools, and decentralized computing techniques in the task of enhancing power system flexibility, security, and reliability [1]. Furthermore, the smart grids paradigm has the potential to support the transformation of existing electrical power systems into dynamic, adaptable, and self-healing networks of distributed and cooperative energy resources.

The suite of smart grids technologies encompasses advanced sensing systems, high-speed communication networks, pervasive monitoring systems, and related tools. These components collectively enable the acquisition of location-specific and real-time actionable data, thereby enabling the provision of enhanced grid services for both system operators, e.g., distribution automation, asset management, and advanced metering infrastructure, and end-users, e.g., demand-side management and demand response [2].

The backbone supporting these technological services is the capacity for heterogeneous and distributed entities, which span from field devices to software processes, to reliably acquire, process and share heterogeneous data. Consequently, the conceptualization of robust and versatile distributed measurement systems emerges as a pivotal concern in design, planning, and operation of smart grids.

In this context, the implementation of wide-area measurement systems (WAMSs) holds the potential for a strategic contribution [3]. WAMSs leverage comprehensive system-wide information to preempt large-scale disruptions and decrease the probability of cascaded faults. This is achieved by enabling adaptive protection and control strategies, which, in turn, aim at enhancing the grid flexibility while mitigating the effects of dynamic perturbations.

The main control and monitoring applications, which could be implemented by a WAMS include [4]

- State estimation: it consists of computing the most probable power system state. In this context, time-synchronized WAMS data, such as bus voltage phasors (e.g., synchrophasors), allows solving this estimation problem by using a linear model, hence, reducing the overall computing complexity, lowering the required computational times, and enhancing the state estimation accuracy.
- Voltage stability analysis: it requires computing specific key performance indicators for each grid bus, such as the Voltage Stability Load Bus Index, in order to promptly detect potential phenomena that could compromise the power system stability.
- On-line grid parameter estimation and dynamic thermal rating: the most important grid parameters can be promptly assessed by processing the time-synchronized WAMS data. Starting from the estimation of these parameters (in particular, the power line resistances), and considering the physical relation between their values and the conductor temperature, it is possible to infer the conductor temperature, hence, enabling the on-line assessment of the transmission reliability margins (dynamic thermal rating).
- Power system congestion management: achieving optimal dispatching necessitates ensuring that the constraints on the power system transfer capability remain unviolated. In conventional approaches, the nominal transport capability is computed offline, considering worst-case conditions and incorporating significant security margins, often resulting in economic opportunity losses. By utilizing time-synchronized WAMS data, real-time capability can be calculated, enabling a more efficient operation of the power system.
- Fault location: WAMS data processing can enhance the accuracy and reliability of fault location.
- Other interesting WAMS-based applications include:
  - Adaptive protections.
  - System integrity protection schemes.
  - Inter-area oscillation monitoring.
  - Model validation.
  - Smart restoration tools.
  - Early warning system.

The foundation of WAMS is rooted in the acquisition of accurate time-synchronized measurements of the voltage phasors and the grid frequency, which are acquired by a network of synchronized sensors located on different buses within the power system. To this aim, phasor measurement units (PMUs) are currently deployed in the task of computing accurate synchronized information about the voltage and current phasors, frequency, and rate of change of frequency. PMUs synchronization requires a common and highly accurate time reference [5,6].

Embracing PMUs within WAMS grants the capability to accurately monitor power flows in interconnected regions and heavily loaded lines, which offers the potential to reliably increase the grid flexibility, by enabling power system operation closer to its stability margins. Furthermore, PMUs data processing enables the real-time monitoring of the power system dynamic behavior, and the real-time identification of inter-area oscillations. The capacity to promptly recognizing and mitigating the effects of such oscillations could allow system operators to enhance the utilization of the transmission and generation assets. As a resultant outcome, renewable power generators could be operated with greater efficiency, consequently reducing the marginal cost of power generation [7].

Nevertheless, in order to exploit the advantages of synchronized WAMS in modern smart grids, it is imperative to address several unresolved challenges.

Specifically, these systems have conventionally been implemented using computing architectures that are based on client/server paradigms. The key constituents of these architectures include the following: PMUs, phasor data concentrators, application software, and a wide-area communication network (WAN). In particular, PMUs compute phasor data and other electrical parameters, transmitting this amalgamated information to the phasor data concentrators. These concentrators, in turn, collect and dispatch these data to a central monitoring center.

Numerous studies have outlined that this hierarchical monitoring framework might not be able to effectively address the rising complexities of modern smart grids, which are characterized by extensive data exchange and high network complexity [8,9]. The technology constraints, the associated costs, the computing complexity, the need for hardware redundancy, the large network bandwidth, and the unaffordable data storage resources emerge as some of the most prominent limitations of centralized WAMS architectures [10].

Consequently, researchers and designers of high-performance WAMS are reevaluating the design options and the underlying assumptions about the scalability, dependability, heterogeneity, manageability, and system evolution over time [4]. Starting from these lines of investigation, this chapter delineates the pivotal role assumed by cooperative and self-organizing smart sensor networks [11]. Specifically, we analyze the prospect of distributing the processing and synchronization functions of WAMS onto an interconnected network of PMUs endowed with distributed consensus protocols [4].

The deployment of consensus protocols enables the cooperation among distributed PMUs, allowing their time synchronization and global coordination obviating the necessity for explicit point-to-point message exchanges or routing protocols. Rather, it disseminates information across the communication network by iteratively updating the state of each PMU through a weighted average of the states of its neighboring PMUs. At each step, each PMU computes a localized weighted least-squares estimate, which ultimately converges to the global maximum-likelihood solution. By virtue of this characteristic, PMUs can achieve both synchronization of their local data acquisitions and perform computations in a fully decentralized manner. This includes the computation of numerous critical variables that define the actual operation state of the power system, without necessitating a central fusion center to aggregate and process the measurements from all PMUs.

This monitoring framework offers a host of advantages over conventional client-server paradigms, particularly in terms of reduced network bandwidth, decreased computation time, and enhanced extensibility and reconfigurability. These traits collectively render the WAMS architecture highly scalable, self-organizing, and distributed, rendering it an optimal choice for tackling wide-area monitoring in smart grids.

3.1 The role of wide-area measurements in smart grids

Modern information and communications technologies are enabling notable progress in the deployment of intelligent systems for monitoring, protection, and management of electrical grids. A multiple array of cutting-edge technological solutions and advanced strategies for grid protection and control can be integrated into the task of developing novel tools for enhancing smart grids operation. In this context, the extensive implementation of WAMS has garnered recognition as one of the most effective enabling technologies.

3.1.1 Elements of WAMS

WAMSs are designed to assist power system operators in addressing the rising level of complexity of modern electrical grids by enabling the comprehensive integration of pervasive measurement technologies, advanced application tools, and wide-area communication infrastructures. Traditionally conceived as autonomous monolithic tools, WAMSs go beyond the conventional functions of supervisory control and data acquisition (SCADA) systems by delivering real-time situational awareness across extensive geographical areas [12]. This feature allows inferring useful insights into various pivotal aspects of power grid operation, thereby furnishing prompt and reliable decision-making support for power system management, enabling the development of adaptive control and protective systems. Additionally, these insights can be used to build an adaptive knowledge base, which aims at promptly classifying the current operation state of the power system and detecting early-stage faults [1].

The effectiveness of WAMSs hinges on the acquisition of accurate phasor and frequency data obtained from a synchronized sensors network properly distributed throughout the power network. Achieving this necessitates the deployment of a network of time-synchronized phasor measurement units (PMUs) that provide precise information concerning voltage and current phasors, frequency, and rate-of-change-of-frequency, all referenced to a highly accurate shared time source.

Ensuring the accurate operation of PMUs asks for the establishment of a uniform and dependable timing reference, which determines when samples of bus voltages and currents are acquired. This can be achieved by synchronizing these samples to a unified timing reference provided by a synchronizing source, whether internal or external to the PMUs. Crucially, these timing signals must adhere to specific criteria:

- aligned with coordinated universal time;
- continuously accessible at all measurement points throughout the interconnected power system;
- characterized by a degree of reliability, availability, and precision in line with strictly specified requirements.

To fulfill these requirements, the utilization of GPS-based timing signals has been widely deployed for the time synchronization of PMUs. This approach brings forth notable advantages, primarily rooted in the fact that it obviates the necessity for establishing a primary time and time distribution systems. Simultaneously, it confers a host of inherent benefits, particularly in terms of global coverage, and adaptability to evolving network configurations.

Conventional WAMSs configurations adhered to hierarchical architectures, which are structured across multiple network layers and integrate an array of technologies for data acquisition, transmission, concentration, and processing. As detailed described in [7], a typical representation of a WAMS architecture hinges on several core constituents, which include PMUs, phasor data concentrators (PDCs), application software, and communication networks. In particular, phasor data and other relevant grid variables are amalgamated by the PMUs and transmitted across the available communication channels to the phasor data concentrators. Subsequently, these information can be relayed to a second collector layer, namely the "Super PDCs," which are directly connected to the monitoring center where WAMS applications are directly executed, alongside the data storage and sharing functionalities.

The set of applications that can be implemented by WAMS strictly depends on the available PMUs network. In particular, in the presence of a reduced PMUs number, only a partial assessment of the actual operation state can be computed, and the WAMS is frequently employed in the task of performing monitoring applications for specific transmission lines, such as voltage stability analysis, dynamic thermal rating assessment, and inter-area oscillation monitoring. In this case, the PMUs number and their optimal location strictly depend on the specific application. A wider spectrum of WAMS applications can be implemented in the presence of a pervasive PMUs network, which allows inferring in real-time accurate information about the actual power system state, enabling the development of complex wide-area applications, such as grid topological analyses, system state estimation, vulnerability assessment, adaptive protection schemes, early warning systems, and fast restoration tools. In this context, WAMS is no longer deployed only for grid monitoring but also for implementing wide-area protective and control functions, hence, evolving to a wide-area measurements protective and control system (WAMPAC).

Field experiences have shown that the application of WAMSs on interconnected power networks allows system operators to avoid large disturbances and reduce the probability of catastrophic blackouts. Besides, it can support the implementation of adaptive protection and control strategies aimed at increasing transmission network capacity and minimizing wide-area disturbances.

Moreover, it is expected that the large-scale deployment of WAMS and WAMPAC in active power distribution systems could provide the benefit of promptly identifying perturbation phenomena, thereby reducing the risk of severe blackouts. In this context, the implementation of WAMS/WAMPAC allows for the deployment of adaptive protection and control functions, aiming to improve the flexibility and reliability of the distribution grids. However, unlocking these potential advantages within modern smart grids involves addressing various challenging issues.

3.1.2 Current state of WAMPACs

In recent years, the implementation of wide-area monitoring functions has become integral across all Continental Europe Power systems, serving as an additional tool for power system operation and analysis. The current interconnected power system extends beyond Continental Europe to include North African countries like Morocco, Algeria, Tunisia, and Turkey. The ongoing expansion of interconnections, particularly facilitated by European Super-Smart-Grid initiatives like e-Highway 2050, necessitates a comprehensive analysis of the current state of wide-area monitoring systems. This analysis aims to evaluate their security and reliability levels.

The capability for international wide-area measurements is made possible through a dedicated communications network that supports data exchanges among different transmission system operators. In particular, ENTSO-E, which is the European Network of Transmission System Operators, representing 39 electricity transmission system operators from 35 countries across Europe, has established this network, known as the Electronic Highway, which functions as a meshed router network separate from the internet. It connects European transmission system operators, facilitating real-time and non-real-time data exchange between them.

Key wide-area monitoring functions employed in European control rooms include the following [14]:

1. Fault analysis support: The high sampling rate of PMUs enables the monitoring of electromechanical phenomena across wide-area systems. Installing PMU devices throughout the system facilitates the easy localization and clearance of faults.
2. System load monitoring: Time-synchronized and time-stamped data collected by PMUs throughout the Continental European system provide wide-area information on active and reactive power flows.
3. Dynamic line thermal monitoring: Real-time monitoring of line temperatures is possible through the measurement of current and voltage phasors, allowing for the estimation of corresponding line series resistance. This technology contributes to reducing congestions and power curtailment.
4. Power system restoration support tool: TSOs can conduct joint tests for system restoration. During real restoration scenarios, precise knowledge of angle differences between different electrical islands, crucial for system resynchronization, can be exploited.
5. Online dynamic system stability monitoring: PMU data offer valuable indices characterizing the power system stability. Assessing interarea oscillations requires synchronized real-time data from various points within the Continental European interconnected power system.
6. Post-event offline analysis: Analyzing offline historical data allows for identifying general behaviors and future trends of the interconnected power system.
7. Pan-European Data Exchange: Data from PMUs installed at the most representative European electrical transmission buses provides a comprehensive view of the actual power system operation state. The synchronized value of

frequency obtained from the received data proves useful in detecting potential grid separation events and in controlling system frequency and deviations.

In particular, the list of existing WAMS-based real-time applications or algorithms includes the following:

- Time-series visualization of PMU data.
- Dynamic line thermal monitoring.
- Voltage angle monitoring.
- Regional angle monitoring for critical power lines.
- Rate of change of frequency (RoCoF) estimation over variable time-window.
- Island identification/detection.
- Perturbation detection (e.g., line faults, generation loss).
- Disturbance analysis (e.g., impact characterization and location).
- Voltage stability assessment (e.g., computing of the voltage eigenvalues, voltage stability index).
- Detection of power/frequency oscillatory patterns.
- Early warnings of incipient faults.
- Analysis of inter-area modes based on continuous measurements and disturbance-based methods.
- Locating contributions to poorly damped or unstable oscillations.
- Higher frequency sub-synchronous oscillation analysis and early warning of resonance.
- System strength monitoring for voltage support.

In the near future, the following WAMS-based functions will be deployed:

- Advanced spectral analysis.
- Power system inertia monitoring.
- Pole slip detection for synchronous generators.
- Wide-area oscillation control system using incipient oscillation detection algorithms to trigger automatic real-time actions.

The mentioned lists represent only a small subset of potential WAMPAC applications. However, critical applications like wide-area protection and control require thorough evaluations of the security and reliability of WAMPACs before they can serve as substitutes for conventional monitoring, protection, and control paradigms.

3.1.3 Improved synchronization accuracy supporting new WAMPACs functions

A typical PMU device can perform 50–100 measurements per cycle with synchronization accuracy of approximately 1 μs, based on available GPS clocks. This level of accuracy is suitable for analyzing fundamental voltages, currents, and system frequencies, but it may not be sufficient for measuring harmonics, particularly, in terms of harmonic phase angles. Currently, there is no established standard for integrating power quality data into WAMPAC systems, and the processing, exchange, and preparation of recorded harmonic measurements for further analysis remain unclear [13].

Another significant concern in WAMPACs is the growing prevalence of high-voltage direct current (HVDC) networks in the European power system. These networks typically operate at switching frequencies of a few (tens) of kilohertz, generating high-frequency voltage harmonics. The propagation of these harmonics through the power system and their interaction with network impedance and load impedance are not well understood. Of particular concern are new power electronic devices with input capacitive filters (EMI filters), introducing new system resonances and potentially acting as "sinks" for high-frequency harmonic currents, leading to equipment malfunction and damage [13].

These novel interactions, with currently unknown system-wide effects, underscore the need to reevaluate requirements in existing standards and manufacturer specifications for the accuracy and synchronization of PMU devices. This calls for the concept of "super-synchronization of PMUs," aiming to enhance and upgrade their current microsecond-range capabilities to the nanosecond range. This enhancement is crucial, especially when investigating frequency ranges up to 100 kHz.

Additionally, new requirements for stability monitoring should be introduced, taking into account the effects of renewable power generators. This aspect becomes increasingly important for ensuring the stability and reliability of power systems.

3.2 Time synchronization in WAMS

The distinctive feature of WAMS, compared to conventional monitoring systems, is the PMU. Hence, it is essential to provide a detailed description of the PMU architecture to emphasize its close relationship with the requirements for precise and resilient time synchronization. A diagram illustrating the components constituting a PMU is presented in Figure 3.1. The input signal is assumed to originate from the output of a current or voltage instrument transformer [14,37].

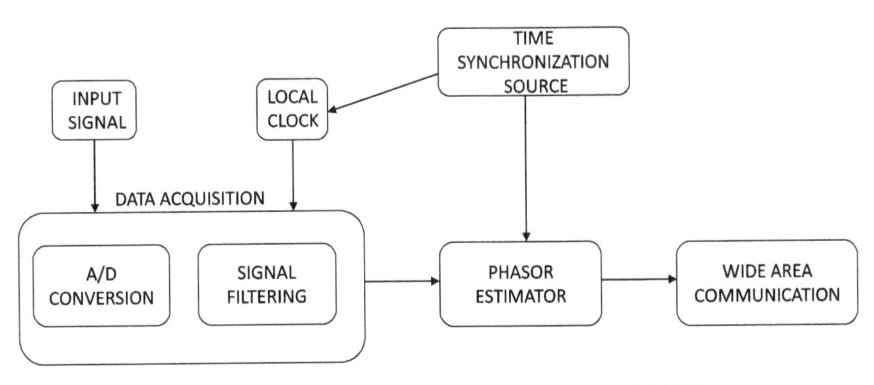

Figure 3.1 Data processing architecture of PMU

- Data acquisition system: This system processes the input signal through a series of components, including a signal conditioner, an anti-aliasing filter, and an analog-to-digital converter, preparing it for subsequent analysis.
- Estimator: This component involves a microprocessor-based unit aimed at calculating the phasor associated with the nominal frequency of the sampled signal. It essentially decomposes the signal using an algorithm that performs discrete Fourier transform.
- Communication module: This module is necessary for transmitting measurements to phasor data concentrators.
- Local clock: PMUs rely on an accurate time synchronization source to measure synchrophasors. While they possess a local quartz oscillator clock for redundancy, it lacks the precision required for calculating synchrophasors. Consequently, PMUs must receive precise time information from an external source.

3.2.1 PMUs synchronization

WAMS can be conceptualized as a network of synchronized sensor nodes, specifically PMUs, distributed within the power systems, where each sensor is equipped with a local clock, which is based on a hardware oscillator. These clocks exhibit distinct physical characteristics and operate under variable ambient conditions.

A comprehensive exploration of key concepts related to time synchronization in sensors network is documented in [15]. The ensuing overview encapsulates the essential elements delineated in the literature.

Each clock endeavors to approximate time as $C(t)$. While an ideal clock would ideally have a rate of change in the approximation $\frac{dC}{dt}$ equal to 1, real clocks contend with clock drift. Clock drift results from unpredictable frequency variations in the oscillators. Consequently, the local clock of a PMU, located at a bus i within the power system, approximates time as

$$C_i(t) = a_i t + b_i \tag{3.1}$$

where a_i and b_i represent the clock drift and the deterministic offset with respect to the reference time, which can be easily corrected.

Comparing two different clocks of the sensor network leads to the following:

$$C_1(t) = a_{12} C_2(t) + b_{12} \tag{3.2}$$

In this case, a_{12} and b_{12} are called relative drift and relative offset, respectively.

Addressing a global synchronization challenge involves aligning $C_i(t)$ for each $i = 1, ..., n$. This alignment can be achieved by equalizing clock rates and offsets or continuously correcting offsets to maintain synchronization over a designated time interval. In the context of WAMS, the latter option is the only viable one due to the stringent accuracy requirements and the wide geographical coverage with varying ambient conditions.

The evaluation of a time synchronization solution performance encompasses several key characteristics [16]:

- Accuracy: it refers to the degree of conformity between the measured time signal and the common reference value.
- Availability: it measures the synchronization architecture ability to disseminate timing services within the required coverage area.
- Continuity: it represents the system capability to guarantee synchronization over fixed time horizon.
- Reliability: it denotes the probability that a synchronization system will fulfill its function within defined performance limits over a specified period under given operating conditions.
- Integrity: it represents the synchronization system ability to detect degradation in timing signals and provide timely warnings to users.
- Coverage: it defines the geographic area where application-specific synchronization system requirements for accuracy, availability, continuity, reliability, integrity, and coverage parameters are concurrently satisfied.

Typical PMUs exhibit a sampling rate of 50/100 Hz, and their synchronization accuracy reaches the level of microseconds, contingent on the latency in GPS communication. While these features are well-suited for analyzing fundamental voltages, currents, and system frequencies, they may not suffice for harmonics measurement. The escalating share of renewable energy sources in Europe and the establishment of a unified European electric market necessitate the creation of new high voltage direct current (HVDC) links between European countries. These HVDC links, operating at high switching frequencies in the tens of kHz, emerge as potent sources of high-frequency voltage harmonics that traverse the system, influencing network and load impedance values.

Additionally, global interest is growing in power system operation under low inertia scenarios. As the penetration of renewable generation rises, system inertia decreases, compromising power system security due to frequency oscillations. This situation has prompted the exploration of new applications, including online inertia estimation. Achieving accurate estimations for such applications, particularly, in scenarios where inertia is considered as an ancillary service, demands high precision.

Moreover, the power system state can be deduced by assessing PMU measurements with consistent timestamps. Furthermore, PMU synchronization is essential for comparing measured phases, which hold significance only in relation to a shared reference time within a very narrow uncertainty range.

The need for accurate time synchronization in PMU can be explained qualitatively by considering the DC power flow equations [17], which allow approximating the active power flow P_{ij} between two buses as

$$P_{ij} = \frac{1}{X_{ij}}(\hat{\delta}_i - \hat{\delta}_j) \tag{3.3}$$

where X_{ij} is the line reactance, and $\hat{\delta}_i, \hat{\delta}_j$ are the voltage angles measured by the PMUs located at the buses i and j, respectively.

As a consequence of synchronization inaccuracy, the measured voltage angle $\hat{\delta}_i$ deviates from the actual voltage angle δ_i by an error $\Delta\delta_i$, which remains independent of the magnitude of the active power flow.

For instance, assuming a time synchronization error $\Delta t_s = 32$ s, the resulting voltage angle error $\Delta\delta_i$ for a power system operating at a frequency $f = 50$ Hz and a corresponding period $T = 0.02$ s are calculated as follows:

$$\Delta\delta_i = \frac{\Delta t_s \cdot 360^\circ}{T} = \frac{32 \cdot 10^{-6} \cdot 360}{0.02} = 0.576^\circ \qquad (3.4)$$

Examining (3.3), it becomes evident that the voltage angle error $\Delta\delta_i$ introduces a proportionate error in the power flow estimation on a power line ΔP_{ij}, strictly dependent on the accuracy of time synchronization. Achieving nano-scale time synchronization accuracy diminishes the phase angle error $\Delta\delta_i$, thereby enhancing the accuracy of power system state estimations.

In contrast to transmission systems, distribution systems exhibit lower voltage angle differences between two buses due to reduced power flows. Consequently, the relative error on the power flow $\frac{\Delta P_{ij}}{P_{ij}}$ attributed to the voltage angle error $\Delta\delta_i$ becomes more pronounced and non-negligible when employing microsecond-scale time synchronization. Therefore, the effective deployment of PMUs in distribution systems, referred to as μPMUs [18], necessitates a synchronization architecture at the nano-second scale.

To ensure compatibility among PMUs, the IEEE Standard for synchrophasors, IEEE C37.118 [19], establishes a minimum accuracy level. The accuracy is gauged through the total vector error (TVE) of a PMU, contingent on both amplitude and phase measurements. Adhering to the Standard necessitates maintaining the TVE below 1%, translating to a maximum time error value of 31.8 s for systems operating at 50 Hz, assuming no amplitude errors. However, considering the contribution of amplitude to TVE, the Standard advocates for an accuracy level on the order of 1 s. This stringent requirement cannot be met by the PMU clock alone, underscoring the necessity for a precise external synchronization source, which is based on atomic clocks.

According to the North American SynchroPhasor Initiative, time synchronization for PMUs should be supplied by a synchronization source, which should be characterized by the following features [20]:

- High accuracy.
- High reliability.
- Wide-area coverage.
- High availability.

Global Navigation Satellite Systems (GNSS), such as the Global Positioning System (GPS), Galileo and Glonass, can provide such features for the time synchronization of PMUs, since they are equipped with atomic clocks.

3.2.2 Characterization of the time synchronization performances

The evaluation of time synchronization performance in WAMS involves measuring the time delay introduced by each PMU clock in relation to an accurate reference clock. This measurement process necessitates three key components [21]:

1. the device under test (DUT), representing the PMU clock being evaluated for accuracy and stability;
2. a reference oscillator, ideally an atomic clock, serving as a highly accurate reference;
3. a time interval counter (TIC), which gauges time offset by receiving start and stop signals from the DUT and the reference oscillator via a pulse per second (PPS) signal [22].

The resultant measurements facilitate the characterization of two primary features:

1. Accuracy: It gauges the deviation of the time measured by the DUT from the reference oscillator, i.e., the time offset. The TIC, equipped with two input signals initiating and halting the counter, measures the delay by counting cycles between the start and stop signals. In cases where measurements follow a normal distribution, accuracy can be described by its mean and standard deviations. However, in the presence of non-stationary measurements, accuracy is time-dependent, requiring additional post-processing analysis [22].
2. Stability: while accuracy assesses the frequency or time offset between two signals, stability indicates whether the time offset remains consistent over time. Stability serves as a statistical estimate of time fluctuations of a signal within a specified time interval.

Stability can be characterized by standard deviation only when dealing with stationary data, where outcomes are time-independent, and noise is white, evenly distributed across the frequency band of measurements. In stationary data, the mean and standard deviation converge to a single value as more data points are added to the measurement. However, real-time measurement data often exhibit non-stationarity, and the mean and standard deviation may not converge to a single value. Consequently, the Allan deviation $\sigma_y(\tau)$ is employed for such scenarios [22].

$$\sigma_y(\tau) = \sqrt{\frac{1}{2(N-2)\tau^2} \sum_{i=1}^{N-2} (x_{i+2} - 2x_{i+1} + x_i)^2} \tag{3.5}$$

In the equation above, x_i represents the ith measurement of the synchronization error, τ denotes the averaging time, which is the time interval considered between two successive measurements in the calculation, and N is the total number of measurements.

Examining an Allan deviation graph with τ on the x-axis and $\sigma_y(\tau)$ on the y-axis provides insights into the optimal averaging time needed to eliminate certain types of noise. The optimum averaging point, known as the "noise floor," is where the remaining noise is attributed to nonstationary processes. This noise floor helps determine the optimal averaging time for revealing the true frequency offset of the DUT [23].

The type of noise can be identified based on the slope of the Allan deviation curve, denoted as α. Common noise types in time and frequency measurements include white phase ($\alpha = 2$), flicker phase ($\alpha = 1$), white frequency ($\alpha = 0$), flicker frequency ($\alpha = -1$), and random walk frequency ($\alpha = -2$). Identifying the noise type is beneficial, as each type requires an optimal statistical approach to handle it [24].

For instance, a white process implies that measurements correspond to a Gaussian random variable, fully characterized by a mean and a standard deviation. This indicates that consecutive observations are randomly distributed and uncorrelated with previous ones. Conversely, flicker and random walk processes exhibit correlations between consecutive observations. A random walk process is characterized by a random distribution of the next observations about the current value, resembling noise superimposed on slowly varying functions. The flicker process lies between random walk and white noise.

Stability can also be characterized in the frequency domain using the power spectral density (PSD), which describes the amplitude of phase (or frequency) fluctuations as a function of Fourier frequency. The PSD aids in evaluating the noise processes associated with a signal, similar to the Allan deviation.

The PSD models the phase variation of a clock using power-law spectral densities of the form:

$$S_y(f) = k(\alpha) f^{\alpha} \tag{3.6}$$

Here, $S_y(f)$ represents the one-sided spectral density of f (Fourier frequency), $k(\alpha)$ is a multiplying coefficient, and α is the exponent of power-law noise processes.

While the analysis in the frequency domain through PSD and time domain through Allan deviation are nearly equivalent, the former is preferred for shorter time intervals, while the latter is suitable for longer ones [25].

3.2.3 Vulnerability analysis of satellite-based WAMSs

The use of satellite-based timing signals for WAMS synchronization has been extensively researched, with various technologies available for PMUs synchronization. In this context, the most commonly used technology is based on GPS signal processing, which is based on a constellation of 24 satellites. Each GPS satellite transmits a spread-spectrum waveform on two carrier frequencies and provides a correction to UTC time, which is automatically applied to the outputs of the receiver. GPS signal processing offers high timing accuracy, limited only by short-term signal reception, with a basic accuracy of 0.2 s.

Another option for synchronized wide-area monitoring is the INMARSAT system satellites, which will carry a GPS-like transponder broadcasting signals that are similar to existing GPS transmissions and can be used with slightly modified GPS receivers.

The European Space Agency's GALILEO system is the third global satellite time and navigation system to come online, comprising a constellation of 30 satellites divided among three circular orbits at an altitude of a certain number of kilometers to cover the Earth's surface. GALILEO provides an integrity signal to ensure the quality of the signals received and to immediately inform the user of any error. The GALILEO time precision in terms of time errors for different signals ranges from 0.7 to 8.1 ns.

Numerous threats can compromise the reliable acquisition of satellite-based time synchronization signals in WAMS systems. This vulnerability can result in the loss or corruption of critical data within a synchrophasor network, which may lead to operational issues in the power system [26].

Indeed, although satellite-based synchronization technologies provide timing accuracy that easily exceeds the requirements of the power industry, they are vulnerable to both intentional or unintentional interferences that could hinder the correct operation of the WAMS. The disruption mechanisms that could trigger these phenomena can be classified into different categories [13]:

- Ionospheric effects: The transit time of satellite signals through the ionosphere can be significantly affected by sunspot activity, which causes an increase in the solar flux, charged particles, and electromagnetic rays emitted by the Sun. As a result, the satellite receiver of the PMU may experience reduced performance in tracking the satellites due to scintillations, which cause rapid variations in the amplitude and phase of the satellite signal. The equatorial and high latitude regions are particularly susceptible to these effects caused by the increased ionospheric activity.
- Unintentional interference: As satellite signals travel through the upper regions of the Earth's atmosphere, they are susceptible to interference from solar disturbances. Additionally, obstructions to the line of sight of satellites, such as in urban areas or near foliage, can cause a considerable decline in the overall quality of the synchronization signal, either temporarily or permanently. Hence, it is imperative to have a dependable estimation of the availability of timing signals.
- Radiofrequency interference: Electronic devices that emit signals in the same frequency band as satellites can cause interference. Although transmitters are usually designed to avoid interfering with wireless signals, faulty or improperly operated equipment can emit signals at the same frequency as satellite signals. This interference can result in poor reception of timing signals by the satellite receiver of the PMU, particularly, if the interference is strong enough.
- Intentional interference: Satellite-based timing signals are vulnerable to radio interference. This interference can cause the signals to weaken and can result in poor reception. Unfortunately, the deliberate jamming of satellite signals is a potential threat to future smart grids. As critical infrastructure, smart grids are vulnerable to external attacks, and intentional interference is considered a serious problem that must be addressed. Despite this, the literature on WAMSs has not extensively explored the potential consequences of cyber-attacks. If the PMUs lose synchronization signals, all the main WAMS functions could be

compromised, jeopardizing the protection and control functions. To mitigate this risk, a comprehensive analysis of potential cyber-attack scenarios is essential.

Consequently, there exist various threats that can impede the proper functioning of satellite-based synchronization technologies and consequently, lead to the loss or corruption of important data in WAMPACs. This can result in operational issues in a power system. As outlined before, the most dangerous threats that could hinder GNSS-based time synchronization are intentional attacks, which can be classified into jamming and spoofing attacks. Jamming disrupts communication between the satellite and the receiver by using higher power signals or by shielding the receiver's antenna. Spoofing, on the other hand, is a more subtle attack that causes the receiver to acquire a counterfeit signal and take it as the true one for the acquisition of precise time and position while being under attack. It is important to note that satellite-based timing signals cannot be considered a trusted input for PMUs on the power system since they are weak in power and do not have any authentication or encryption schemes against cyber-attacks.

It is important to note that power systems can suffer significant financial losses as a result of these cyber-attacks, which can be carried out using simple and inexpensive electronic devices. Despite the severity of these attacks, there is still much research to be done in understanding their impact on power systems. One of the challenges is detecting GPS spoofing, which can go unnoticed by certain detection algorithms.

In this context, one of the most severe threats is the time synchronization attack, which is a spoofing attack consisting in a two-step process: initiating a disturbance that interrupts the receiver ability to maintain GPS signal and transmitting a false GPS signal, which is then received in place of the genuine one. This attack is extremely difficult to detect, since it can be implemented by using inexpensive and widely available software and hardware tools and does not require the physical access to PMUs [27].

3.2.4 Time synchronization attack in WAMS

The time synchronization process of WAMS has been designed to ensure precision and reliability. However, it is susceptible to cyber-attacks. GPS-based time synchronization is vulnerable to a variety of malicious attacks, as discussed in a recent publication [28]. These attacks can be categorized as GPS jamming and spoofing. Jamming is when a GPS signal is deliberately blocked from reaching a PMU, while spoofing involves sending a false GPS signal to the PMU, resulting in the failure of time synchronization.

Although time synchronization process is designed in such a way that precision and reliability are guaranteed, it has almost no defense against cyber-attacks. Unfortunately, time synchronization through GPS is vulnerable to a wide range of malicious attacks, as described in [28], which can be broadly classified as GPS jamming and spoofing. Jamming is characterized by directly blocking the GPS signal from reaching a PMU, whereas spoofing consists in sending to the PMU a fake GPS signal, causing the failure of time synchronization.

These phenomena make modern power systems extremely vulnerable to undetectable cyber-attacks that can bypass bad-data detection algorithms. This is because most WAMS currently installed in power grids are not designed to be resilient to cyber-attacks.

Unobservable attacks pose a huge threat to grid operations, especially with the increasing reliance on automatic operations based on data from monitoring devices. One type of invisible threat to PMUs is the time synchronization attack (TSA) [27].

A TSA is a spoofing attack that can cause the loss of synchronization, corrupt time-stamps and phase angle measurements in a PMU. The attack is based on GPS spoofing, where a malicious attacker introduces a fake GPS signal that causes an offset in the receiver's clock, which in turn affects the PMU phase angle measurements and timestamping process.

The attack can be easily carried out using universal software radio peripheral devices, which are cheap and widely available [29]. A TSA does not require physical access to the PMUs and is difficult to detect. The attack is usually carried out in two steps:

1. the attacker launches an interference that causes the receiver to lose track of the GPS signal;
2. the attacker launches a counterfeit GPS signal with a higher signal-to-noise ratio, which will be acquired instead of the true signal.

The TSA has different impacts on power systems, as several studies have shown [21,23,30]. Indeed, it has been demonstrated that TSA can hinder secure and reliable power system operation, affecting almost all applications related to WAMPACs.

A thorough evaluation of all the potential detrimental effects of TSA on power systems is presented in [30], which emphasizes that monitoring applications can be misled by the false information gathered by corrupted PMUs, leading to incorrect corrective actions. Protection applications can also be affected, as a TSA can cause a fault location error larger than 100 km, which can significantly reduce the reliability of the entire system.

In particular, consider the scenario where a GPS receiver in a PMU is subjected to a spoofing attack. This could result in inaccurate phase measurements, leading to several issues. First, applications that depend on phase angle measurements could be affected, as they would receive misleading information. This could lead to false corrective actions, both manual and automatic, that might cause problems for the systems.

Studies have reported that power flow estimation errors could be as high as a hundred MW [31]. This could cause economic losses due to inefficient generation dispatch, even if catastrophic events do not occur. Second, protection schemes could be affected as well, leading to islanding of part of the distributed generation from the power systems.

In a highly renewable generation scenario, this could cause massive cascade faults due to frequency variations. Additionally, PMU systems that are used to detect fault location could be impacted, as spoofing attacks could provide false information about the fault position. This could delay the clearing of the fault, especially if a maintenance team needs to physically reach the location. Studies have estimated that location errors could be as large as 180 km.

Moreover, many protection schemes rely on the measurement of phase angles or frequency, such as in anti-islanding procedures. However, during a TSA, some of these devices may erroneously trip, resulting in negative impacts on the power network. Additionally, accurate measurements are also crucial for controlling power system stabilizers, and coordinated damping schemes. Many studies (e.g., [32]) highlighted that TSA may even cause instability in power systems.

Finally, GPS spoofing could also cause delays in feedback control loops, which could degrade the controllers performance. To tackle these issues, a new research area, the real-time rejection and mitigation of time synchronization attacks, has emerged.

The literature suggests various approaches to mitigate the effects of TSAs, including detection and mitigation as well as protection. While some techniques, such as state estimation proposed by [33], focus on detecting TSA in a single PMU, most detection methods primarily address balanced three-phase transmission systems, except for the approach taken by [34] which targets unbalanced systems.

Protection mechanisms rely on satellite signal encryption and authentication, as suggested by [35]. However, implementing these solutions requires modifications to the monitoring infrastructure or PMUs and fails to address other critical aspects of satellite signals for time synchronization. For instance, power systems, being one of the most critical infrastructure, cannot solely rely on satellite signals because of their uncertain availability and data integrity.

Therefore, the development of accurate and dependable synchronization sources is of utmost importance to enhance the cyber-resilience of conventional satellite-based systems, particularly, in the context of WAMS.

3.3 Emerging needs in wide-area smart grids monitoring

Even though WAMSs have been available in the market since the early 1990s, the exploration of their potential applications within modern smart grids is still in its initial stages. A specific issue in this context involves deploying such systems using reliable and highly scalable architectures, which should be resilient against multiple contingencies that could perturb their operation and, in turn, impact the dependable functioning of the smart grids.

In finer detail, as previously outlined, the deployment of WAMS has followed client–server-based architectures, with PMUs time synchronization frequently achieved through satellite signal processing.

Numerous studies [10] questioned about the inherent limitations of this hierarchical monitoring setup, which seems potentially unfeasible in the context of future smart grids, where the increasing complexity of the power network and the demand for pervasive grid monitoring ask for more flexible and scalable computing frameworks.

Indeed, it is expected that the advent of the smart grid will amplify current data collection by approximately four orders of magnitude, causing the centralized WAMS architectures to swiftly reach saturation points. Consequently, the data streams received from distributed PMUs might not allow system operators with the essential insights required to promptly respond within the short timeframes required to effectively mitigate the impact of large disturbances. Even in the presence of accurate models designed to translate data into actionable information, smart grid operators are confronted with the subsequent challenges:

- communication bottlenecks;
- computing complexities;
- need for reducing the monitoring time period;
- growing complexity of energy management systems;
- resilience of the computing facilities to external attacks.

Additionally, the reliance on satellite-based synchronization hinges on the transfer of information through air communication channels. The wireless characteristic of satellite communication links, coupled with the limited power levels of satellite signals, makes them susceptible to radiofrequency interference. Indeed, any source emitting electromagnetic radiation has the potential to disrupt satellite signals if it emits radio signals within the frequency bands used by satellites. Hence, the pursuit of dependable and efficient synchronization techniques, with the aim of furnishing redundant timing signals for PMUs, represents a relevant issue to address in the task of ensuring the accurate and secure operation of WAMSs. These backup time synchronization sources become operational when the satellite signals are unavailable, furnishing a more dependable timing source. Moreover, even if one signal experiences degradation or becomes unavailable, PMUs should continue operating according to strict quality constraints.

As these challenges are confronted and tackled, researchers and WAMS designers are reevaluating a multitude of aspects tied to design, reliability, heterogeneity, manageability, and the system evolution over time [3,8,9]. These studies suggest a transition toward a more decentralized architecture for WAMSs. To meet these requisites, the most promising technologies facilitating such advancements comprise:

- The deployment of new monitoring paradigms aimed at spreading computing intelligence at each level of the WAMS architecture.
- The evolution of conventional wide-area communication networks to pervasive systems, which not only allow the PMUs to exchange data with the control centers but also to share data with other PMUs and with all the power components installed in the substation.

However, the definition of a more effective WAMS architecture within the context of smart grids is still in its early stages and necessitates advancements in both theoretical frameworks and practical implementations.

In our perspective, solving this matter entails finding an optimal balance between a cost-efficient solution with enhanced management capabilities and heightened performance and reliability. Particularly, noteworthy is the fact that a centralized WAMS concentrates intelligence at a singular central location, simplifying the monitoring process maintenance and affording greater control to system operators. Nonetheless, a drawback to a fully centralized monitoring model is the need to collect all data back to a remote application function, which introduces a potential single point of failure for the network. Furthermore, this might lead to slow-down the communications when dealing with substantial volumes of raw data.

On the flip side, decentralized WAMS configurations empower local PMUs to make independent decisions without relying on communication with a central "master" device. This has the potential to enhance response times and eliminate single points of failure. However, the implementation of intelligent PMUs capable of autonomous decision-making demands increased processing power and more extensive upkeep. Hence, the cost associated with a completely decentralized WAMS architecture could surpass those of a fully centralized network.

Our firm conviction is that the most effective WAMS architecture should be a fusion that synergistically integrates these two monitoring paradigms. Guided by such a vision, this chapter describes the potential role of cooperative and self-organizing smart sensor networks equipped with distributed consensus protocols within WAMSs. Specifically, we delve into the feasibility of decentralizing WAMS processing and synchronization functions across a network of interactive PMUs equipped with distributed consensus protocols.

The employment of consensus protocols empowers PMUs to reach an agreement on global grid information, enabling their coordinated collaboration. This decentralized paradigm spreads information throughout the communication network by updating each PMUs state based on a weighted average of its neighboring states. At every iteration, each PMU computes a local weighted least-squares estimate, converging toward the global maximum-likelihood solution.

3.4 A self-organizing WAMS architecture

Within this section, we describe a WAMS architecture based on the "think locally act globally" approach. The core idea is to exploit the principles of information spreading in the task of coordinating a network of collaborative PMUs [36,37]. Each individual PMU holds the capacity to (i) gather local bus measurements by interacting with an array of sensors and (ii) establish communication with a restricted group of neighboring PMUs.

The implementation of WAMS processing and synchronization functions relies on the PMUs ability to exchange and process local information through distributed consensus algorithms. This is achieved through the integration of a dynamic system (oscillator) within each PMU, which is initialized with the local data, and whose evolution is ruled by the local coupling strategies described in Chapter 2. This bio-inspired framework allows all PMUs to reach a consensus on general functions of the variables sensed by the entire PMUs network.

This property empowers local PMUs to attain time synchronization in data acquisition, enabling each PMU to compute critical variables defining the actual power system operation state without necessitating a central fusion center. Consequently, the fundamental functions of WAMS can be executed by the PMUs in a fully decentralized and non-hierarchical manner. This characteristic bestows the proposed architecture with self-organizing, highly scalable, cooperative, and distributed attributes, rendering it an optimal option for smart grids monitoring.

3.4.1 Synchronization functions

The synchronization of PMU clocks stands as a crucial cornerstone of WAMSs, which enable the distributed clock systems to accurately sample the local grid data according to a common time scale.

The information propagation algorithms founded on the interdependence of dynamic systems, as elaborated in Chapter 2, represent a viable alternative to the conventional centralized and hierarchical clock synchronization strategies. In essence, PMUs attain time synchronization by adjusting their clocks in accordance with the local coupling strategies, bypassing the necessity for any cluster header. Indeed, every PMU acts as a source of time synchronization signals to other PMUs. Once this network of interconnected oscillator clocks achieves synchronization, each clock will display an identical value, which remains constant once reached, hence, empowering the inherent PMU oscillators to be locked to a common and unique time reference, despite variations in individual oscillators frequencies.

The efficacy of this decentralized synchronization paradigm has undergone evaluation across diverse research domains, spanning from the realm of biology to technology-driven fields.

3.4.2 Monitoring functions

Implementing the collaborative framework described in Chapter 2 enables the PMUs to reach a consensus on the principal variables defining the actual operation of the smart grid, hence, allowing each PMUs network to perform the monitoring functions of WAMSs.

Specifically, when PMUs sense the bus voltage magnitudes, the subsequent array of observations can be employed to initiate the dynamic systems:

$$\omega_i = (V_i, |V_i - V^*|) \tag{3.7}$$

where V^* and V_i are the nominal and the measured voltage magnitude at the ith bus, respectively. This choice allows the PMUs network to synchronize with the average grid voltage magnitude and the mean voltage magnitude deviation, namely

$$\omega^* = \left(\frac{\sum_{i=1}^{N} V_i}{N}, \frac{\sum_{i=1}^{N} |V_i - V^*|}{N} \right) \tag{3.8}$$

where N is the buses number.

Other relevant variables describing the actual grid operation state can be inferred by a different choice of the observation vector. For example, if the PMUs measure the active and reactive power injected at the monitored bus, the following observation vector could be defined:

$$\omega_i = (N\,(P_{Gi} - P_{Di}), N\,c_{pi}(P_{Gi})P_{Gi}, N\,c_{qi}(Q_{Gi})Q_{Gi}) \tag{3.9}$$

where

- P_{Gi} and P_{Di} are the active power generated and demanded at the ith bus;
- Q_{Gi} is the reactive power injected at the ith bus;

- c_{pi} and c_{qi} are the incremental costs of the active and reactive powers generated at the ith bus.

This choice of the observation vector allows the PMUs to reach a consensus on the active power losses and the overall cost of the generated active and reactive power:

$$\omega^* = \left(\sum_{i=1}^{N} (P_{Gi} - P_{Di}), \sum_{i=1}^{N} c_{pi}(P_{Gi})P_{Gi}, \sum_{i=1}^{N} c_{qi}(Q_{Gi})Q_{Gi} \right) \tag{3.10}$$

Further useful variables, such as power quality indices, can be readily evaluated through an alternative definition of the observation vector.

Through the adoption of this biologically-inspired framework, every PMU can both assess the variables of the monitored bus (gathered via the embedded sensors) and infer the main variables describing the actual state of the entire smart grid (evaluated by inquiring about the built-in dynamical system state). This dual insight empowers a continuous comparison between local and global metrics, enabling responsive measures to be implemented if the bus features significantly deviate from the current grid performance.

3.5 Simulation studies

In this section, we analyze the utilization of the described self-organizing WAMS architecture for the purpose of synchronized grid monitoring for IEEE 118-bus test system. This particular power network is composed of 118 buses, 186 branches, 91 load buses, and 54 generators, as detailed in [4].

A network of 118 cooperative time-synchronized sensors has been deployed across the power network, with each sensor assigned to a respective bus. The coupling coefficients ruling the local coupling between the built-in oscillators have been defined according to the same adjacency matrix characterizing the power network.

3.5.1 Monitoring functions

Utilizing the described decentralized monitoring approach enables individual sensors to calculate both the performance metrics of the monitored bus through the acquisition of localized data and the grid performance metrics achieved via localized information exchange with neighboring sensors. This framework empowers the system operator to evaluate the core variables defining the real-time grid operation by querying any sensor, obviating the necessity for a central fusion center to aggregate and process all the measured data.

Specifically, we assume that each cooperative sensor acquires the following bus variables: voltage magnitude, generated and demanded active power. Hence, the observation array is structured as follows:

$$\omega_i = \left(V_i, N\,(P_{Gi} - P_{Di}), (V_i - V^*)^2\right) \tag{3.11}$$

This choice allows the sensors to reach a consensus on the average grid voltage magnitude, the active power losses, and the average voltage magnitude deviation:

$$\omega^* = \left(\frac{\sum_{i=1}^{N} V_i}{N}, \sum_{i=1}^{N} (P_{Gi} - P_{Di}), \frac{\sum_{i=1}^{N} (V_i - V^*)^2}{N} \right) \tag{3.12}$$

The corresponding trajectories of the state derivatives $\dot{\theta}_i$ of each dynamic agent for a fixed grid operating state are depicted in Figures 3.2 and 3.4. In these figures, it has been assumed a linear coupling function with a system gain of 0.2.

Moreover, the decentralized solution of the maximum/minimum consensus problems allows the dynamic agents to promptly synchronize to the max/min value of the grid variables, as shown in Figures 3.5 and 3.6, which report the trajectories of the state derivatives $\dot{\theta}_i$ in the task of computing the minimum and maximum bus voltage magnitudes.

As expected, the implementation of distributed consensus protocols allows the convergence of all the dynamic agents (initiated with random initial states) toward the actual values of the grid variables. The network synchronization is achieved within approximately 350 s for the computation of active power losses, and merely 0.5 s for determining the maximum and minimum bus voltage magnitudes. The larger convergence time necessary for active power loss computation can be attributed to the greater complexity inherent in solving the average consensus problem. Notably, it is worth mentioning that the incorporation of a nonlinear coupling protocol is expected to sensibly reduce the required convergence iterations.

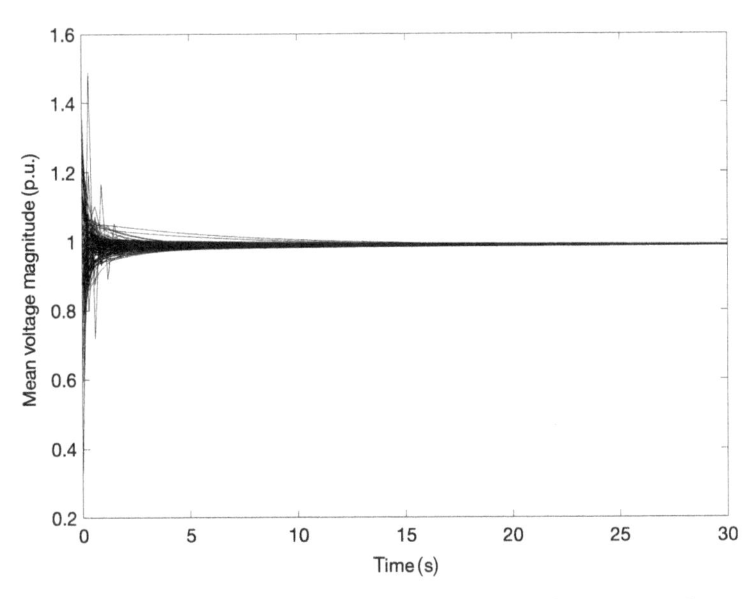

Figure 3.2 Evolution of the state derivatives in the task of computing the average grid voltage magnitude

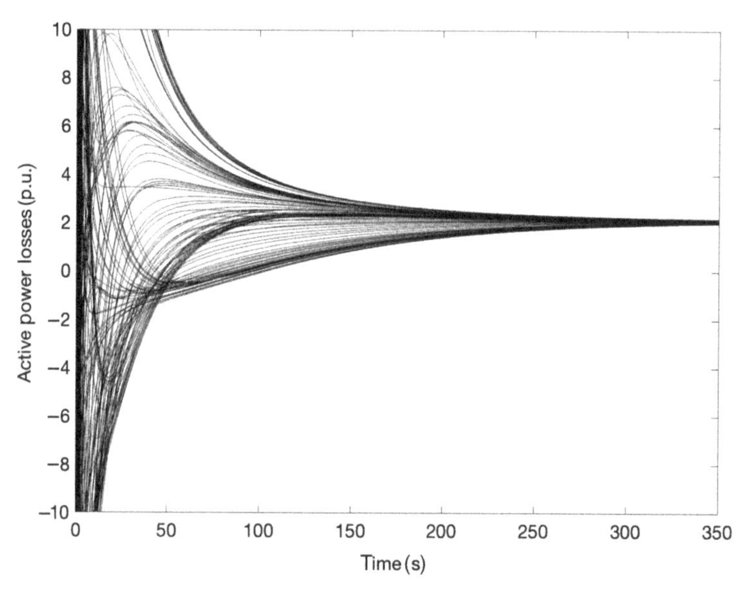

Figure 3.3 Evolution of the state derivatives in the task of computing the active power losses

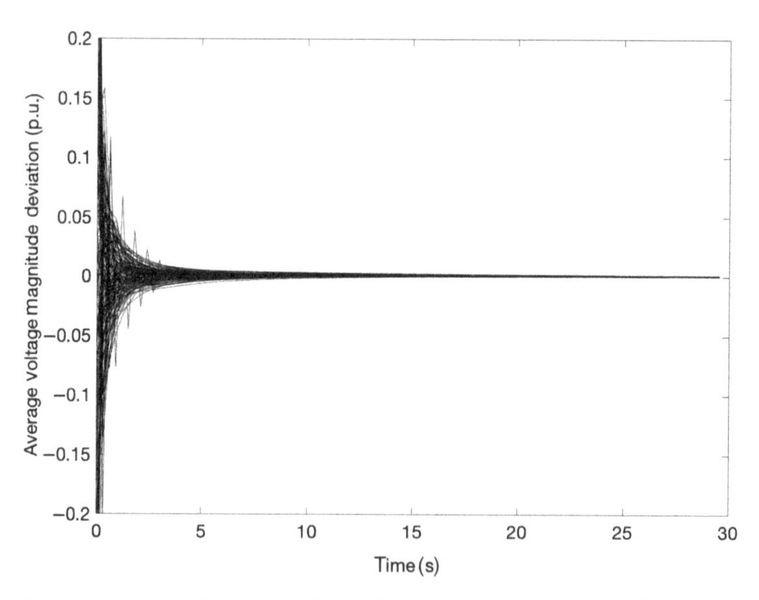

Figure 3.4 Evolution of the state derivatives in the task of computing the average voltage magnitude deviation

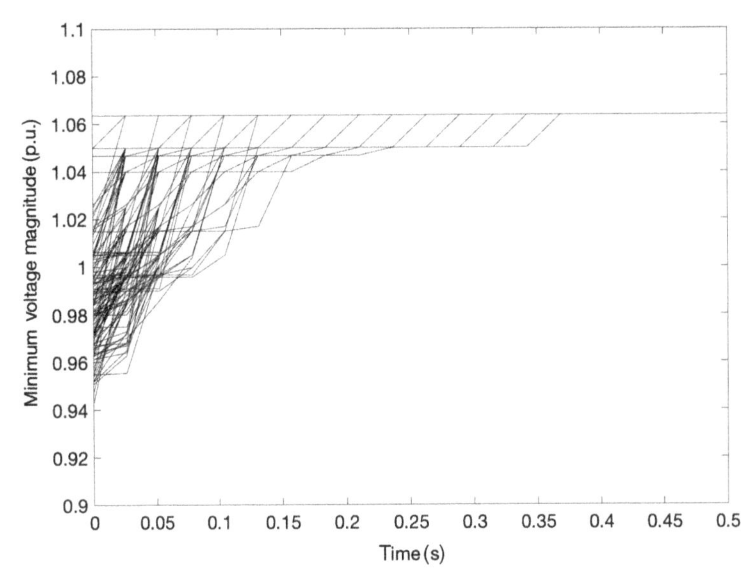

Figure 3.5 Evolution of the state derivatives in the task of computing the maximum bus voltage magnitude

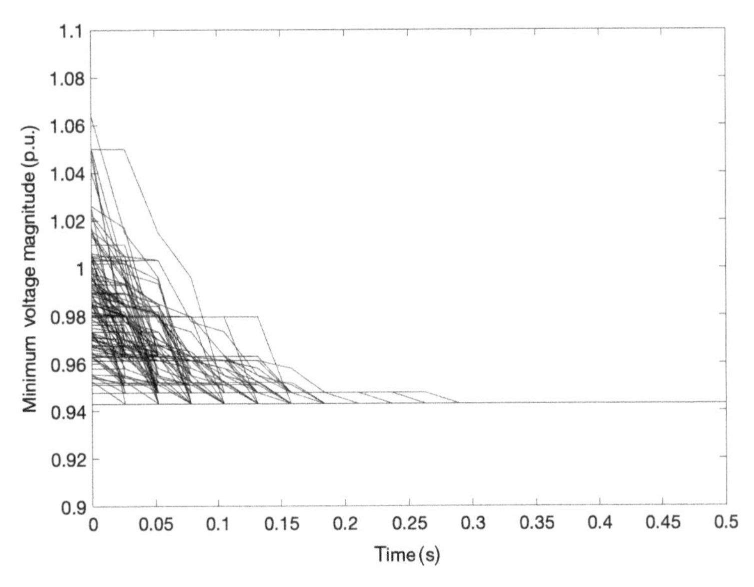

Figure 3.6 Evolution of the state derivatives in the task of computing the minimum bus voltage magnitude

3.5.2 Synchronizing functions

To validate the efficiency of the decentralized paradigm in sensors time synchronization, we conducted examinations of both linear and nonlinear coupling strategies. To facilitate this investigation, we evaluated the subsequent functions:

$$h(.) = \begin{cases} x \text{ Linear} \\ \frac{\exp Lx - 1}{\exp Lx + 1} \text{ Exp} \\ \sin\left(2\pi \frac{x}{L}\right) \text{ Sin} \end{cases} \tag{3.13}$$

When evaluating the time synchronization performance, we focused on determining the number of iterations necessary to achieve a synchronization error below 10^{-4} s. The outcomes of this assessment are summarized in Table 3.1.

Table 3.1 Local coupling parameters

Coupling mode	**K**	**L**	**Iterations**
Type 1	0.1	–	283
Type 2	0.1	3	274
Type 3	1.2	50	192

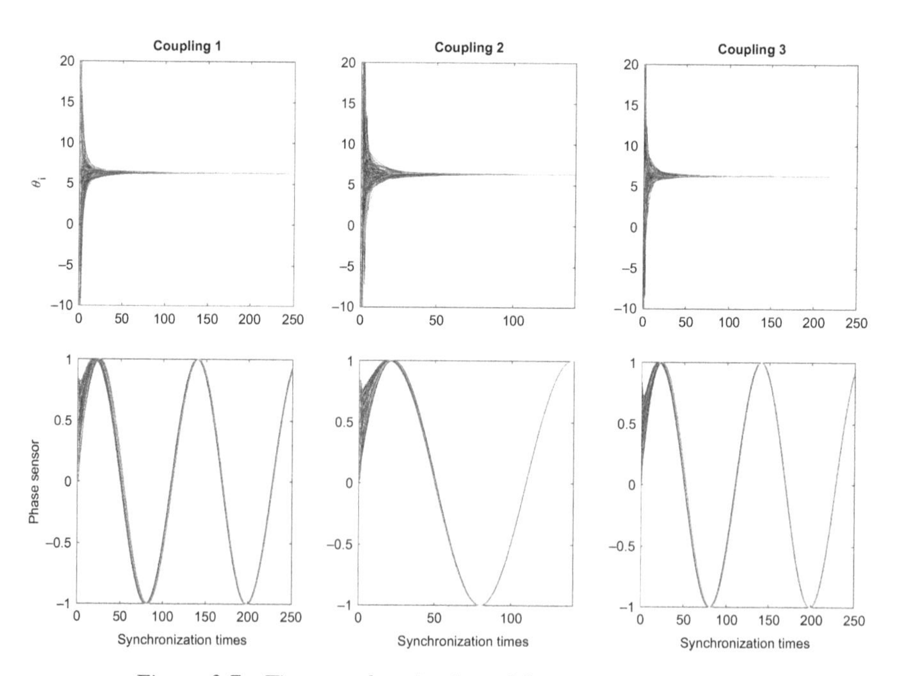

Figure 3.7 Time synchronization of the cooperative sensors

Upon analyzing this data, it is notable that the utilization of a nonlinear coupling strategy yielded a significant reduction in convergence times. This feature becomes more apparent when examining the dynamic evolution of the dynamic agents state, as depicted in Figure 3.7.

These findings provide validation for the efficiency of the described paradigm in time-synchronization of the sensors data acquisition. This feature could be regarded as an additional synchronization source, serving as a backup in instances where GPS signals are unavailable or degraded.

References

[1] Terzija V, Cai D, Valverde G, *et al.* Flexible wide area monitoring, protection and control applications in future power networks. In: *10th IET International Conference on Developments in Power System Protection (DPSP 2010). Managing the Change*; 2010. p. 1–5.
[2] He H. Toward a smart grid: integration of computational intelligence into power grid. In: *The 2010 International Joint Conference on Neural Networks (IJCNN)*; 2010. p. 1–6.
[3] IEEE Smart Grid Vision for Computing: 2030 and Beyond Reference Model. 2016. p. 1–18.
[4] Vaccaro A, Iacoviello A, and Popov M. Cooperative and self organizing sensor networks: the enabler for smarter grids. In: *Sensor Networks for Sustainable Development.*
[5] Chakrabarti S, Kyriakides E, Bi T, *et al.* Measurements get together. *IEEE Power and Energy Magazine*. 2009;7(1):41–49.
[6] Hu Y and Novosel D. Challenges in implementing a large-scale PMU system. In: *2006 International Conference on Power System Technology*; 2006. p. 1–7.
[7] Terzija V, Valverde G, Cai D, *et al.* Wide-area monitoring, protection, and control of future electric power networks. *Proceedings of the IEEE*. 2011;99(1):80–93.
[8] Madani V and King RL. Strategies to meet grid challenges for safety and reliability. *International Journal of Reliability and Safety*. 2008;2(1):1–11.
[9] Giri J, Parashar M, Trehern J, *et al.* The situation room: control center analytics for enhanced situational awareness. *IEEE Power and Energy Magazine*. 2012;10(5):24–39.
[10] Yang Q, Barria JA, and Green TC. Communication infrastructures for distributed control of power distribution networks. *IEEE Transactions on Industrial Informatics*. 2011;7(2):316–327.
[11] EL Brak M, EL Brak S, Essaaidi M, *et al.* Wireless sensor network applications in smart grid. In: *2014 International Renewable and Sustainable Energy Conference (IRSEC)*; 2014. p. 587–592.
[12] Popelka A, Jurik D, Marvan P, *et al.* Advanced applications of WAMS. In: *22nd International Conference and Exhibition on Electricity Distribution (CIRED 2013)*; 2013. p. 1–4.

[13] Varriale E, Morante Q, Vaccaro A, *et al.* Development and experimental validation of a super-synchronized phasor measurement unit. In: *2019 International Symposium on Advanced Electrical and Communication Technologies (ISAECT)*; 2019. p. 1–5.
[14] Pepiciello A. Flexibility solutions for the integration of variable renewable energy sources in power systems. *PhD Thesis*, University of Sannio; 2021.
[15] Sivrikaya F and Yener B. Time synchronization in sensor networks: a survey. *IEEE Network*. 2004;18(4):45–50.
[16] Vaccaro A, Zobaa AF, and Formato G. Vulnerability analysis of satellite-based synchronized smart grids monitoring systems. *Electric Power Components and Systems*. 2014;42(3-4):408–417.
[17] Purchala K, Meeus L, Van Dommelen D, *et al.* Usefulness of DC power flow for active power flow analysis. In: *IEEE Power Engineering Society General Meeting*, 2005. IEEE; 2005. p. 454–459.
[18] Shahsavari A, Farajollahi M, Stewart E, *et al.* Situational awareness in distribution grid using micro-PMU data: a machine learning approach. *IEEE Transactions on Smart Grid*. 2019;10(6):6167–6177.
[19] IEEE. IEEE Standard for Synchrophasor Measurements for Power Systems; 2011.
[20] Force NTST. Time Synchronization in the Electric Power System. Technical Report, North American Synchrophasor Initiative; 2017.
[21] Pepiciello A, Vaccaro A, and Pietropaoli T. Experimental assessment of a PTP-based system for large scale time synchronization of smart grids. In: *2020 55th International Universities Power Engineering Conference (UPEC)*; 2020. p. 1–6.
[22] Lombardi MA. Fundamentals of time and frequency. *The Mechatronics Handbook*. 2002, CRC Press, Boca Raton.
[23] Pepiciello A, Vaccaro A, and Pietropaoli T. Enabling technologies for resilient time synchronization of wide area power system monitoring. In: *Encyclopedia of Electrical and Electronic Power Engineering*. Amsterdam: Elsevier; 2023. p. 88–98.
[24] Kent A and Williams JG. *Encyclopedia of Computer Science and Technology*. Vol. 45-Supplement 30. London: CRC Press; 2002.
[25] Riley WJ. *Handbook of Frequency Stability Analysis*. US Department of Commerce, National Institute of Standards and Technology; 2008.
[26] Vaccaro A, Zobaa AF, and Formato G. Vulnerability analysis of satellite-based synchronized smart grids monitoring systems. *Electric Power Components and Systems*. 2014;42(3–4):408–417.
[27] Humphreys TE, Ledvina BM, Psiaki ML, *et al.* Assessing the spoofing threat: development of a portable GPS civilian spoofer. In: *Proceedings of the 21st International Technical Meeting of the Satellite Division of The Institute of Navigation (ION GNSS 2008)*; 2008. p. 2314–2325.
[28] Zhang H, Peng S, Liu L, *et al.* Review on GPS spoofing-based time synchronisation attack on power system. *IET Generation, Transmission & Distribution*. 2020;14(20):4301–4309.

[29] Humphreys TE, Ledvina BM, Psiaki ML, *et al.* Assessing the spoofing threat: development of a portable GPS civilian spoofer. In: *Proceedings of the ION GNSS International Technical Meeting of the Satellite Division*. Vol. 55; 2008. p. 56.
[30] Almas MS, Vanfretti L, Singh RS, *et al.* Vulnerability of synchrophasor based WAMPAC applications to time synchronization spoofing. *IEEE Transactions on Smart Grid*. 2017;9(5):4601–4612.
[31] Zhang Z, Gong S, Dimitrovski AD, *et al.* Time synchronization attack in smart grid: impact and analysis. *IEEE Transactions on Smart Grid*. 2013;4(1):87–98.
[32] Bi T, Guo J, Xu K, *et al.* The impact of time synchronization deviation on the performance of synchrophasor measurements and wide area damping control. *IEEE Transactions on Smart Grid*. 2016;8(4):1545–1552.
[33] Fan X, Du L, and Duan D. Synchrophasor data correction under GPS spoofing attack: a state estimation-based approach. *IEEE Transactions on Smart Grid*. 2017;9(5):4538–4546.
[34] Delcourt M, Shereen E, Dán G, *et al.* Time-synchronization attack detection in unbalanced three-phase systems. *IEEE Transactions on Smart Grid*. 2021;12(5):4460–4470.
[35] Bhamdipati S, Mina TY, and Gao GX. GPS time authentication against spoofing via a network of receivers for power systems. In: *2018 IEEE/ION Position, Location and Navigation Symposium (PLANS)*; 2018. p. 1485–1491.
[36] Cortés J. Distributed algorithms for reaching consensus on general functions. *Automatica*. 2008;44(3):726–737.
[37] Delvenne JC, Carli R, and Zampieri S. Optimal strategies in the average consensus problem. In: *2007 46th IEEE Conference on Decision and Control*; 2007. p. 2498–2503.

Chapter 4

Decentralized voltage regulation by self-organizing controllers network

The power system decarbonization programs, the modern environmental policies, and the large-scale development of the deregulated energy markets serve as catalysts for the extensive implementation of smart grid technologies. This emerging paradigm could hold a strategic position in propelling the transformation of conventional power grids into active, adaptable, and self-healing energy networks interlinking distributed and cooperative energy resources [1–3].

Advancements in smart grid research have the potential to amplify the efficiency of modern electrical power systems by bolstering the widespread integration of small-scale distributed generation systems, facilitating the exploitation of renewable energy sources, reducing greenhouse gas emissions, and enhancing the dependability of electricity delivery to consumers.

Within this context, a pivotal concern lies in augmenting the integration of renewable power generators into smart grids through effective coordination of their functioning, while mitigating their side effects on grid operations and control.

It is indeed a well-established fact that the integration of renewable power generators into power grids perturbs the power flows and the voltage profiles for both customer and utility equipment, leading to various ancillary consequences (e.g., bidirectional power flows and heightened fault current levels [4,5]). These phenomena heighten the intricacies involved in operating electricity distribution systems, particularly in several European nations and rural settings, where systems are typically structured for radial operation devoid of customer-side generation.

In this context, effective grid voltage regulation is a pivotal challenge that necessitates attention [6,7]. Voltage regulation in electrical grids typically relies on the optimal coordination of multiple control devices, all oriented toward bolstering load bus voltages and enhancing power quality at the distribution level.

With the incorporation of renewable power generators into electrical grids, the complexity of the regulation process intensifies. The injection of active power from these sources carries inherent randomness, thereby influencing the voltage magnitude profiles of all the network buses. This underscores the requirement for a well-designed approach aimed at ensuring efficient grid operation through the proper coordination of the renewable power generators operation with the traditional voltage regulation devices. A crucial consideration here is maintaining a low computational burden to enable effective real-time solutions to the voltage regulation problem.

In addressing these demands, nonlinear programming techniques for optimal grid voltage regulation in the presence of distributed generators have been proposed in [6–8]. These methodologies advocate centralized control strategies for computing the optimal set points of the voltage regulators for each network state. These set points are identified through the minimization of a composite cost function, accounting for a blend of control objectives (such as voltage profile optimization, power loss reduction, and minimization of reactive power costs), while adhering to various constraints (like reliability metrics, voltage stability thresholds, etc.).

However, it has been recognized in recent works that such a hierarchical control paradigm bears inherent drawbacks, particularly in terms of limited scalability, due to the growing pervasiveness of renewable power generators. This expansion could impose unsustainable computational demands and necessitate hardware redundancy, as central processing units become vulnerable single points of failure [9]. Additionally, this approach calls for significant network bandwidth and substantial data storage resources [10].

These challenges pose potential impediments to the applicability of centralized voltage regulation structures in smart grids, especially considering the constant escalation of grid complexity and the need for extensive renewable power generator integration. In this context, current trends lean toward advanced regulation architectures that depart from the older centralized model, moving toward distributed, field-based systems with an increased deployment of intelligent devices, such as smart sensors, where central controllers play a limited role [11,12]. The integration of smart sensors into smart grid voltage regulation can lead to a more efficient distribution of tasks among the voltage controllers, consequently lightening the load on centralized computing facilities and enhancing the efficiency of the voltage regulation system.

Drawing from the trajectories laid out by these works, this chapter seeks to describe a decentralized, non-hierarchical voltage regulation architecture based on intelligent and cooperative smart entities.

The main idea is to deploy distributed voltage controllers, which aim at regulating the voltage magnitude of the grid buses equipped with renewable power generators. Each controller is comprised of a cluster of sensors capturing local bus variables (such as voltage magnitude and active/reactive bus power) and a dynamic agent (oscillator) initiated by the sensor computations. The oscillators of neighboring controllers are interlinked through the local coupling strategies described in Chapter 2. This biologically inspired approach facilitates the prompt synchronization of all the dynamic agents to the weighted average of variables sensed by the voltage controllers [13,14]. Consequently, each voltage controller can independently assess significant variables defining the global smart grid operation (e.g., mean grid voltage magnitude and power losses), without necessitating a central fusion center to collect and process all node measurements.

The knowledge of these global variables allows ascertaining the optimal configuration of the distributed voltage controllers. To this end, each voltage controller processes global variables to achieve two key objectives:

1. assess the evolution of an objective function defining the voltage regulation goals;

2. determine appropriate control actions aimed at minimizing this objective function by regulating the reactive power flows injected by the renewable power generators into the electrical grid.

In this context, we describe a decentralized architecture based on the self-synchronization of the dynamic agents for assessing the cost function and its gradient, and three distributed optimization strategies aimed at identifying the optimal configuration of the voltage controllers. In particular, we analyze the decentralized solution of the voltage regulation problem by a gradient-based minimization algorithm, a meta-heuristic technique based on simulated-annealing, and a fuzzy-based solution technique.

The described voltage regulation strategies present several advantages over traditional client–server paradigms, such as reduced network bandwidth requirements, quicker computation times, and ease of extension and reconfiguration. These attributes make the decentralized architectures highly scalable, self-organizing, and distributed.

4.1 Problem formulation

The primary objective of the voltage regulation procedure is to determine the optimum value of the regulating devices, which enhance the voltage quality by minimizing the bus voltage magnitude deviations from the nominal value and the active power losses.

In existing distribution systems, this objective is typically realized through by coordinating the operation of underload tap-changing transformers and remotely controlled capacitor banks in the task of reducing the bus voltage magnitude deviations and enhancing the voltage quality at the load buses.

Beyond these conventional devices, an augmented integration of distribution flexible AC transmission system (DFACTS) components, such as thyristor tap-changing transformers, dynamic voltage restorers, distribution static synchronous compensators, and power electronic transformers, is expected in modern smart grids.

The significant proliferation of distributed generation sources in smart grids introduces heightened complexity into this regulation process, driven by two key reasons [1].

- In cases where distributed generators are operating in a voltage following mode, namely they are directly linked to the electrical grid, operating at a near-constant power factor, the inherent randomness of power injections can perturb the voltage profiles across all the power networks. This, in turn, necessitates an ongoing fine-tuning of the voltage regulators [5].
- Alternatively, when distributed generators function in a voltage support mode, namely their network connection is realized via a power electronic interface, which enables them to serve as controllable sources of reactive power, they introduce additional degrees of freedom in the solution of the voltage regulation problem. These additional degrees of freedom can be harnessed to notably enhance the voltage profile throughout the distribution network [6].

As a result, effective voltage regulation within a smart grid necessitates the application of an adaptive methodology capable of orchestrating the multitude of available voltage-regulating devices. This orchestration should ensure secure and cost-efficient grid operation, along with the optimal coordination between distributed generators operating in voltage support mode and other conventional voltage regulating devices. Moreover, this methodology should demand minimal computational effort to enable practical online solutions to the voltage regulation problem. This is of paramount importance in managing the inherent uncertainties associated with renewable power generators functioning in voltage-following mode.

In broad terms, solving the voltage regulation problem entails determining the optimal set-points of the voltage regulation devices that, for each grid state, minimize fixed cost functions, and satisfy a set of equality and inequality constraints. Given the presence of both discrete and continuous variables, the overall problem can be formulated as a mixed-integer nonlinear programming problem.

The configuration of the voltage regulation devices is defined by the subsequent vector:

$$y = [Q_{dg,1}, \ldots, Q_{dg,N_g}, Q_{cap,1}, \ldots, Q_{cap,N_c}, V_{F,1}, \ldots, V_{F,N_f}, m] \tag{4.1}$$

where

- $Q_{dg,i}$ is the reactive power injected by the ith $(i = 1, \ldots, N_g)$ distributed generator operating in voltage-supporting mode;
- $Q_{cap,j}$ is the reactive power generated by the jth $(j = 1, \ldots, N_c)$ capacitor bank;
- $V_{F,k}$ is the set-point of the kth $(k = 1, \ldots, N_f)$ DFACTS;
- m is the tap position of the tap-changing transformer.

Note that the vector y takes values in the solution space Ω:

$$y \in \Omega \Longleftrightarrow \begin{cases} tap_{\min} \leq m \leq tap_{\max} \\ Q_{dg,\min,i} \leq Q_{dg,i} \leq Q_{dg,\max,i} \ \ \forall i \in [1, N_g] \\ Q_{cap,\min,j} \leq Q_{cap,j} \leq Q_{cap,\max,j} \ \ \forall j \in [1, N_c] \\ V_{F,\min,k} \leq V_{F,k} \leq V_{F,\max,k} \ \ \forall k \in [1, N_f] \end{cases} \tag{4.2}$$

The solution to the voltage regulation problem should satisfy the strictly technical constraints required for secure grid operation, such as the allowed voltage magnitude ranges (i.e., $V_{\min,i} \leq V_i \leq V_{\max,i} \ \forall i \in [1, n]$) and the maximum allowable line current magnitudes for the N_L distribution lines (i.e., $I_l \leq I_{max,l} \forall l \in [1, N_L]$).

The cost functions describing the voltage regulation objectives may describe both technical and economic figure of merits. This is conventionally represented through a weighted sum of N_o normalized design objectives:

$$f_{opt} = \sum_{i=1}^{N_o} \alpha_i \frac{F_i}{\bar{F}_i}$$

where α_i and $\bar{F}_i$ represent the weighting factor and the normalization threshold of the ith design objective F_i, respectively.

The minimization of the following objectives is frequently considered in solving the voltage regulation problem:

1. active power losses:

$$F_1 = P_G - P_D \tag{4.3}$$

where P_G and P_D are the total active power generated and demanded in the power system, respectively;

2. average voltage magnitude deviation:

$$F_2 = \frac{\sum_{i=1}^{N} (V_i - V_i^*)^2}{n} \tag{4.4}$$

where V_i and V_i^* are the current and the nominal voltage magnitude at the ith bus, respectively, and N is the buses number;

3. maximum voltage magnitude deviation:

$$F_3 = \max_i (|V_i - V_i^*|) \tag{4.5}$$

4. regulation cost for the time period Δt:

$$F_4 = \left(F_1 c_{loss} + \sum_{i=1}^{N_g} c_{dg,i} Q_{dg,i} + \sum_{j=1}^{N_c} c_{cap,j} Q_{cap,j} + \sum_{k=1}^{N_f} c_{F,k} Q_{F,k} + c_{QT} Q_T \right) \Delta t \tag{4.6}$$

where

- c_{loss} is the marginal cost of the active power;
- $Q_T \Delta t$ and c_{QT} are the reactive energy absorbed from the transmission system and the corresponding incremental cost, respectively;
- $c_{dg,i}$ and $c_{cap,j}$ are the marginal costs of the reactive energy generated by the ith distributed generator and j^{th} capacitor bank, respectively;
- $Q_{F,k} \Delta t$ and $c_{F,k}$ are the reactive energy generated by the DFACTS devices and the corresponding marginal cost, respectively.

Given the inherent competition among the design objectives, the solution of the voltage regulation problem is not unique, and the identification of a suitable compromise among these objectives should be obtained.

4.2 A decentralized gradient-descent-based solution of the voltage regulation problem

A decentralized/nonhierarchal architecture for voltage regulation in smart grids can be defined by considering a network of self-organizing controllers, each one controlling the voltage magnitude of the local electrical bus [13,14]. The basic elements composing these decentralized voltage controllers include

- an array of conventional grid sensors aimed at sensing the main bus variables, such as the bus voltage magnitude, and the active and reactive power injected into the grid;
- an array of dynamic oscillators, which are initialized by a proper combination of the bus measurements, and their state is coupled to the states of their neighbors by the strategies discussed in Chapter 2.

- a short-range communication system, which allows the voltage controllers to share their oscillators state;
- a local optimizer, which aims at controlling the reactive power flow locally generated by the available reactive power source.

We will see that the adoption of this architecture allows solving the voltage regulation problem for the actual power system operation state in a fully decentralized way. Moreover, it allows the decentralized controllers to compute the main variables describing the voltage grid performance, and the objective function of the voltage regulation problem.

In particular, if the voltage controllers acquire the bus voltage magnitude, and their dynamic oscillators are initialized by the following observation vectors:

$$\omega_i = [V_i, |V_i - V_i^*|] \tag{4.7}$$

Then, if the controllers reach a consensus, all their oscillators self-synchronize to the average value of the grid voltage magnitude and the voltage magnitude deviation:

$$\dot{\theta}_i = \left[\frac{\sum_{i=1}^n V_i}{n}, \frac{\sum_{i=1}^n |V_i - V_i^*|}{n}\right] \tag{4.8}$$

While, if the following observation vector is adopted to initialize the oscillators:

$$\omega_i = [n(P_{G_i} - P_{D_i}), nc_{P_i}P_{G_i}, nc_{Q_i}Q_{G_i}] \tag{4.9}$$

Where P_{G_i} and P_{D_i} are the active power generated and demanded, Q_{G_i} is the reactive power generated, and c_{P_i} and c_{Q_i} are the incremental costs of the active and reactive powers, respectively.

Then all the oscillators self- synchronize to the active grid losses, and to the active and reactive power production costs:

$$\dot{\theta}_i = \left[\sum_{i=1}^n P_{G_i} - P_{D_i}, \sum_{i=1}^n c_{P_i}P_{G_i}, \sum_{i=1}^n c_{Q_i}Q_{G_i}\right] \tag{4.10}$$

Due to the deployment of this biologically inspired concept, every voltage controller is aware of both the parameters that define the monitored bus (detected through integrated sensors) and the overall variables characterizing the real-time performance of the entire power grid (inferred by analyzing the state of the dynamic oscillators).

The knowledge of these variables enables each voltage controller to evaluate the actual values of the cost functions $(F1, ..., F4)$ and to implement appropriate regulation strategies aimed at minimizing these functions. As described in the next sections, this matter is tackled by designing local optimizers, which aim at regulating the reactive power flows generated by the available voltage regulators.

4.2.1 Local optimizer

The reactive power flows generated at the *i*th bus are regulated by the self-organizing controllers according to a decentralized gradient descent-based solution algorithm.

Initially, the regulation algorithm undertakes the estimation of the variables characterizing the current power system operation state, by self-synchronization of the oscillators network. Subsequently, these variables are processed in the task of inferring $f_{opt,k}$ and $\Delta f_{opt,k}$, which, respectively, denote the value and derivative of the objective function at the specific time point k.

Once this evaluation of variables is accomplished, the variation of the ith control parameter at time step k (referred to as Δy_k^i) can be inferred by employing a gradient descent technique [15]:

$$\Delta y_k^i = -\mu \frac{\Delta f_{opt,k}}{y_k^i - y_{k-1}^i} \tag{4.11}$$

Where μ is a design parameter governing both the stability and the convergence of the algorithm. In this context, it is worth observing that the manipulation of discrete control variables necessitates the deployment of specialized strategies aimed at modifying the iteration procedure formalized in (4.11), which encompass outer approximation techniques, branch-and-bound methods, extended cutting plane methods, and generalized Bender's decomposition [1,16]. These strategies typically hinge on consecutively solving coupled nonlinear programming problems.

Despite its intrinsic limitations, which encompass the need for controllers network to compute the gradient of the cost function, and its tendency to converge to local minima, decentralized gradient descent stands out as a favorable optimization approach for solving the voltage regulation problem, since it exhibits conceptual simplicity and is characterized by robust stability and rapid convergence. Indeed, the iteration scheme formalized in (4.11) enables voltage controllers to progress toward a local minimum of the objective function by taking steps proportionate to the negative approximation of its gradient at the present time step. The incorporation of inequality constraints into the solution process can be achieved using a modified version of the conventional gradient descent algorithm [17].

Upon determining Δy_k^i, the updated set of regulation mechanisms is recognized, validated, and applied to the grid. This iterative procedure concludes upon reaching a predetermined stopping criterion (i.e., the absolute alteration in the objective function falls below a preset tolerance). In such instances, no further optimization is required, and the controllers network solely computes the real-time value of the objective function. Should a variation in the objective function be detected by the network (owing to alterations in the grid state), the network detects the need for computing a new optimal solution, and a minor perturbation of the present set of regulation devices is implemented on the grid to acquire an initial numerical estimation of the gradient. The algorithm then progresses following the aforementioned phases.

4.2.2 Example

This section discusses the deployment of the described decentralized solution scheme in the task of solving the voltage regulation problem for the IEEE 30-bus test system [1].

As far as the available voltage regulators are concerned, six distributed generators operating in grid-supporting mode and characterized by the data reported in

Table 4.1 Characteristics of the distributed generators

Bus	Q_{min} (MVar)	Q_{max} (MVar)
10	−10	10
16	−10	10
20	−10	10
23	−10	10
29	−10	10
30	−10	10

Table 4.1 have been considered in this study [1]. The corresponding reactive power generated are the state variables of the voltage regulation problem, namely:

$$y = [Q_{dg,1}, Q_{dg,2}, Q_{dg,3}, Q_{dg,4}, Q_{dg,5}, Q_{dg,6}] \tag{4.12}$$

We assume that each state variable is regulated by a voltage controller and a network of 30 cooperative sensors has been deployed across the power system (with one sensor corresponding to each bus) in the task of inferring the actual grid operation state. The interconnections between the sensors, described by the coupling coefficients a_{ij}, have been assumed according to the connectivity matrix that defines the electrical network topology. Moreover, the objective function under consideration is

$$f_{opt} = \alpha_1 \frac{\frac{1}{30}\sum_{i=1}^{30} V_i}{M_1} + \alpha_2 \frac{\sum_{i=1}^{30} (P_{G_i} - P_{D_i})}{M_2} + \alpha_3 \frac{\frac{1}{30}\sum_{i=1}^{30} (V_i - V_i^*)^2}{M_3} \tag{4.13}$$

which is formulated as a weighted combination of three normalized objectives: the average magnitude of the grid voltage, the active power losses, and the average voltage magnitude deviation. It is important to note that even more general objectives could potentially be included and incorporated into equation. This selection, however, does not undermine the effectiveness of the described voltage regulation framework.

The subsequent values have been adopted for the weighting factors and thresholds for normalization:

$$\begin{cases} \alpha_1 = \alpha_2 = \alpha_3 = 1 \\ M_1 = 1 \\ M_2 = 0.05 \\ M_3 = 4 * 10^{-4} \end{cases} \tag{4.14}$$

For evaluating the actual value of the objective function (4.13), every node sensor acquires the bus voltage magnitude, as well as the active and reactive bus power. These specific observations are then arranged into the subsequent vector:

$$\omega_i = [V_i, P_{G_i} - P_{D_i}, (V_i - V_i^*)^2] \tag{4.15}$$

Table 4.2 Initial network state

Bus	P_{G_i} (p.u.)	P_{D_i} (p.u.)	V_i (p.u.)
1	0.8223	0	1.0200
2	0.3000	0.2170	1.0000
3	0	0.0240	1.0022
4	0	0.0760	0.9977
5	0.3000	0.9420	1.0000
6	0	0	0.9967
7	0	0.2280	0.9901
8	0.3000	0.3000	1.0000
9	0	0	0.9794
10	0	−0.0580	0.9608
11	0.3000	0	1.0000
12	0	0.1120	0.9782
13	0.3000	0	1.0000
14	0	0.0620	0.9624
15	0	0.0820	0.9597
16	0	−0.0350	0.9691
17	0	0.0900	0.9575
18	0	0.0320	0.9491
19	0	0.0950	0.9461
20	0	−0.0220	0.9505
21	0	0.1750	0.9475
22	0	0	0.9482
23	0	−0.0320	0.9526
24	0	0.0870	0.9387
25	0	0	0.9453
26	0	0.0350	0.9437
27	0	0	0.9502
28	0	0	0.9987
29	0	−0.0240	0.9633
30	0	−0.1060	0.9722

and the following synchronization states can be reached by the dynamic oscillators:

$$\dot{\theta}^* = \left[\frac{1}{30} \sum_{i=1}^{30} V_i, \sum_{i=1}^{30} P_{G_i} - P_{D_i}, \frac{1}{30} \sum_{i=1}^{30} (V_i - V_i^*)^2 \right] \tag{4.16}$$

The knowledge of these synchronization states, and the characteristic data reported in (4.14), allows the voltage controllers to compute the actual value of objective function (4.13). The corresponding trajectories of the oscillators state derivative for the considered power system operation state are reported in Figures 4.1–4.4.

Examining the obtained data reveals that dynamical oscillators (initialized with random values) achieve synchronization in approximately 500 iterations. As a result, the complete time required for synchronization (encompassing factors such as data latency, data re-transmission, and imperfect data reception) amounts to roughly 50 s

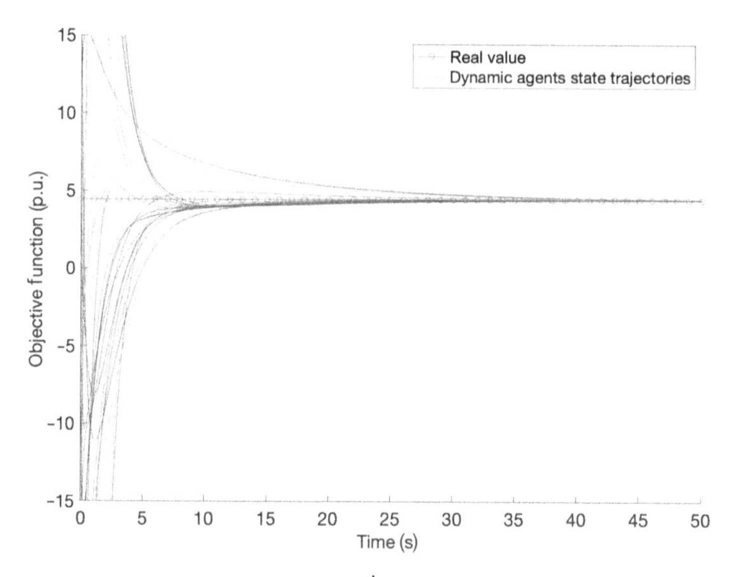

Figure 4.1 *State derivative trajectories ($\dot{\theta}_i$) self-synchronizing on the actual value of the objective function f_{obj}*

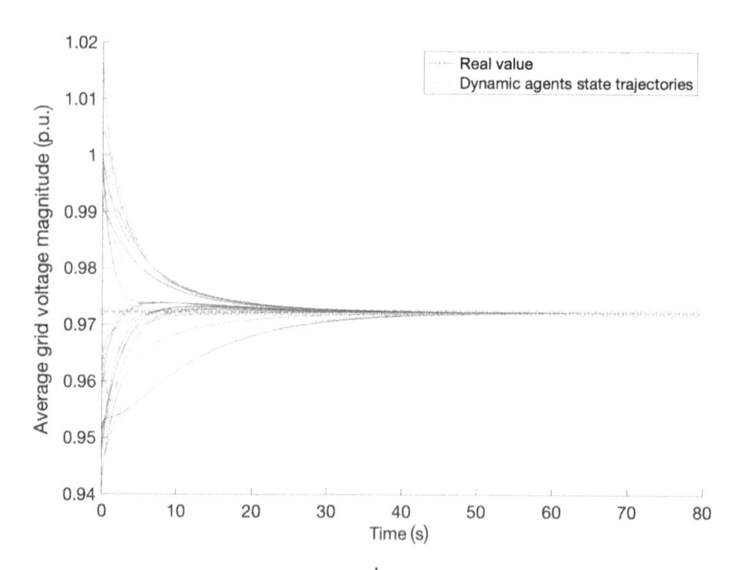

Figure 4.2 *State derivative trajectories ($\dot{\theta}_i$) self-synchronizing on the actual value of the average grid voltage magnitude*

for wireless communications. This aligns well with the time constraints of the voltage regulation processes in Smart Grids.

Starting from this initial grid state, local optimizers were activated in the task of computing the optimal set-points of the voltage regulators. The outcomes obtained are depicted in Figures 4.5 and 4.6.

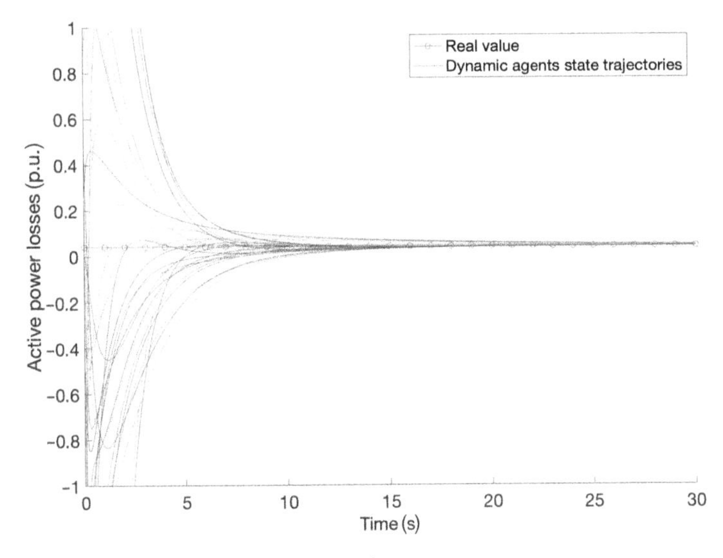

Figure 4.3 State derivative trajectories ($\dot{\theta}_i$) self-synchronizing on the actual value of the active power losses

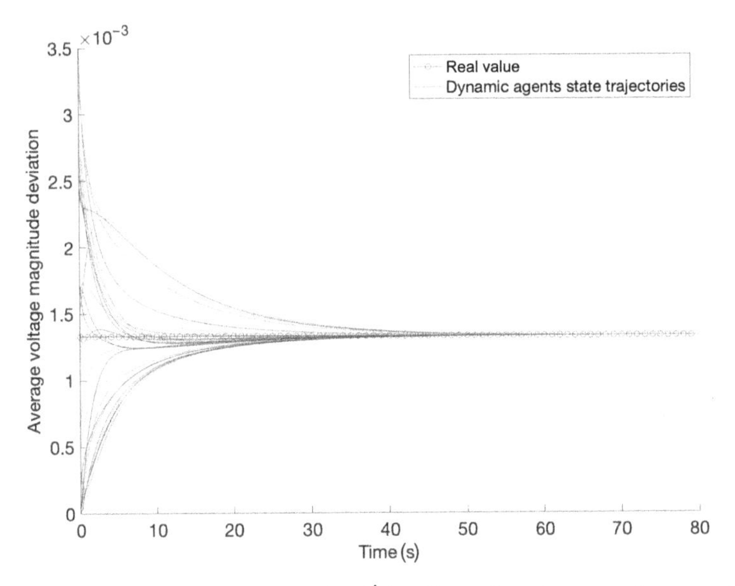

Figure 4.4 State derivative trajectories ($\dot{\theta}_i$) self-synchronizing on the actual value of the average voltage magnitude deviation

Upon reviewing these results, it is useful highlighting that the decentralized voltage regulation strategy results in a significant enhancement of the objective function within approximately 4 min.

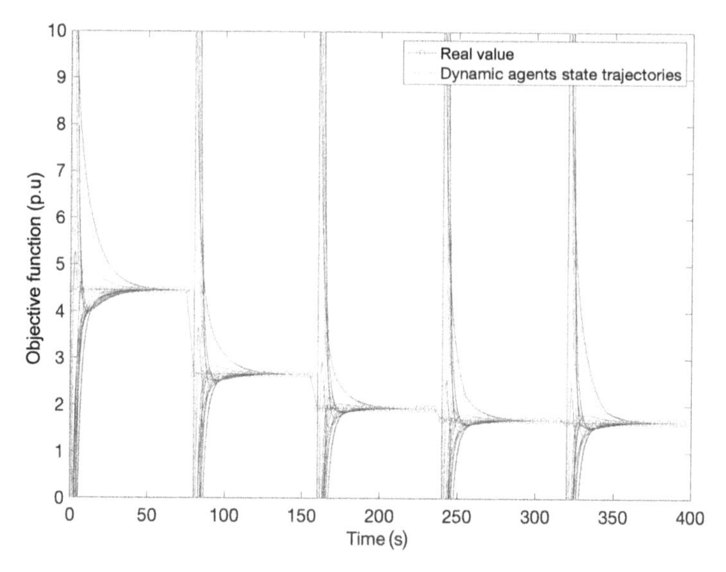

Figure 4.5 State derivative trajectories ($\dot{\theta}_i$) self-synchronizing on the actual value of the objective function

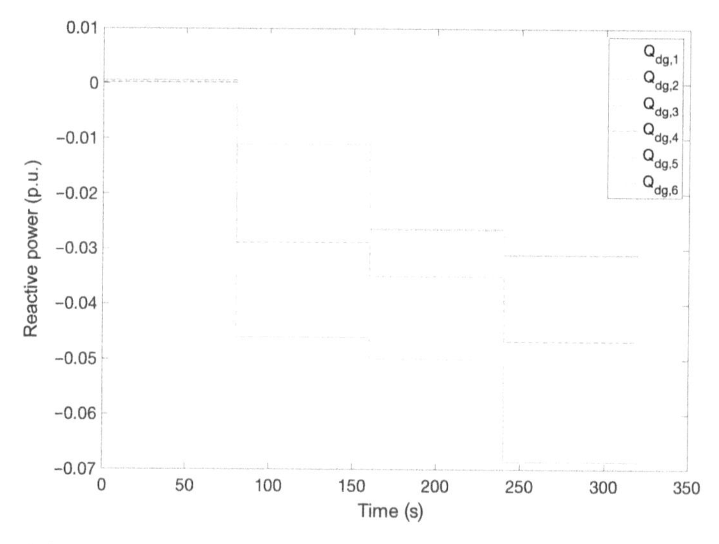

Figure 4.6 Set-points of the voltage regulators identified by the local optimizers

This assertion finds further validation through an examination of the corresponding progression of average grid voltage magnitude, power losses, and mean voltage magnitude deviation, as illustrated in Figures 4.7–4.9.

These figure of merits describe the overall voltage performance of the electrical grid, and as such, they have been taken into account to quantify the benefits deriving by the deployment of the described decentralized regulation strategy. Specifically,

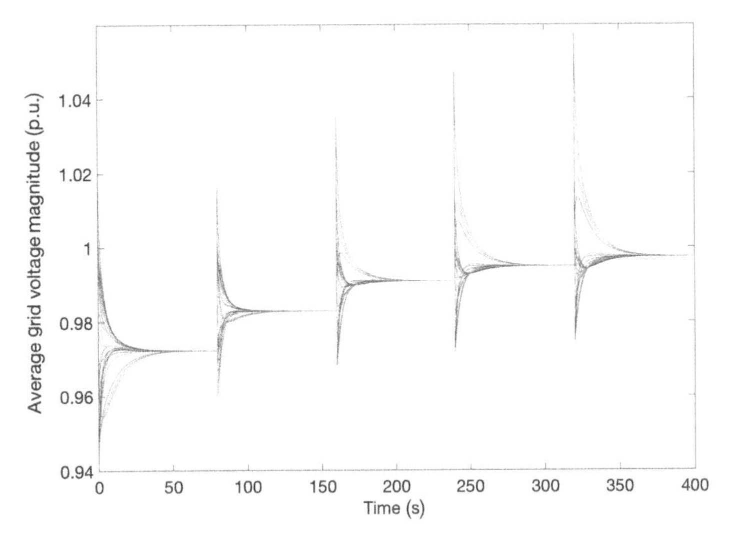

Figure 4.7 *State derivative trajectories ($\dot{\theta}_i$) self-synchronizing on the actual value of the average grid voltage magnitude*

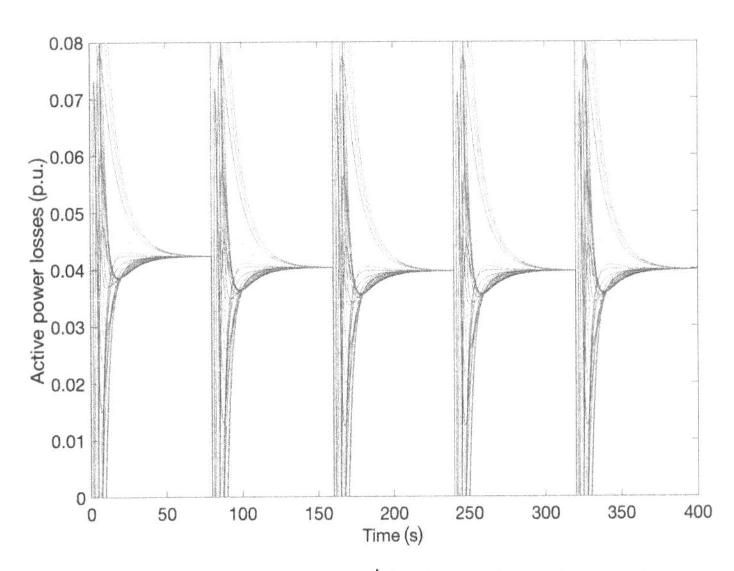

Figure 4.8 *State derivative trajectories ($\dot{\theta}_i$) self-synchronizing on the actual value of the active power losses*

the mean grid voltage magnitude converges toward the designated value (1 p.u.), alongside a reduction in power losses and a diminished average voltage magnitude deviation.

To define a reference point for evaluating the effectiveness of the described decentralized methodology, the optimal regulation problem has been solved by using

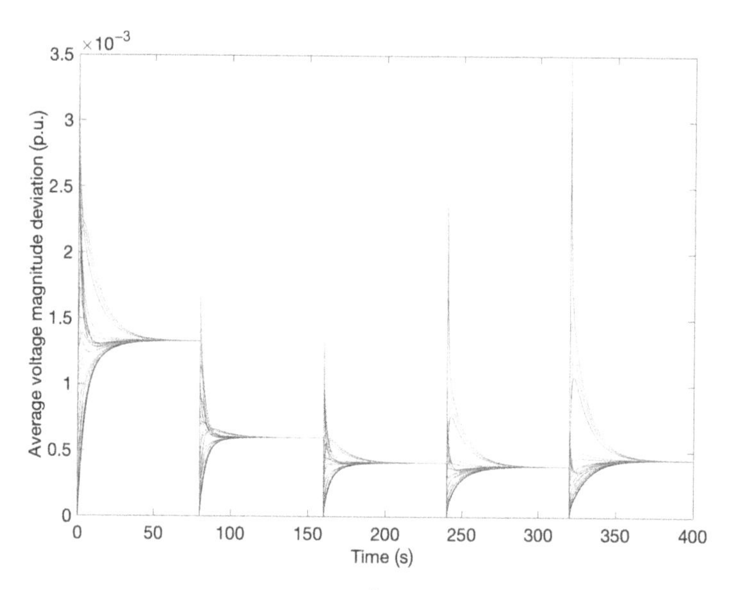

Figure 4.9 State derivative trajectories ($\dot{\theta}_i$) self-synchronizing on the actual value of the average voltage magnitude

a centralized solution algorithm relying on the interior point method. The outcomes obtained are documented as $f_{opt} = 1.6557$ compared to 0.9494.

The simulated outcomes confirm that the solution derived from the distributed voltage controllers is worse than those obtained by the centralized solution attained via the robust optimization algorithm. As anticipated, the interior point-based optimization technique showcases superior performance concerning the minimization of the cost function. However, this method asks for a detailed power network model and a central fusion center responsible for collecting and processing all sensor measurements.

In contrast, the described method identifies a sub-optimal solution to the voltage regulation problem by employing a fully decentralized and non-hierarchical approach for information processing. Indeed, the actual values of the cost function and its gradient are evaluated without relying on a fusion center. Additionally, the decentralized regulation strategy is ruled by distributed optimizers that process both global and local variables, which makes this framework scalable, self-organizing, and distributed.

4.3 A decentralized metaheuristic-based solution of the voltage regulation problem

The local optimizers, which control the reactive power flows injected into the electrical grid by the voltage regulators, can be implemented by using a decentralized search minimization method that relies on simulated annealing (SA) [18]. In this context, after self-synchronizing on the global variables that define the current state of the power system, the cooperative voltage controllers compute the value of the objective function denoted as $f_{opt,k}$, and adjourn the regulator set-points, denoted as yk, by using a search technique based on simulated annealing [16]. The underlying principle of this algorithm involves treating $fopt,k(yk)$ as descriptive of an energy function $E(yk)$. This approach entails generating a new potential solution, $yk+1$, in close proximity to the current solution, yk, in a randomized manner. Acceptance of the new solution hinges on whether it leads to a decrease in the objective function value (i.e., if $f_{opt,k} > f_{opt,k+1}$). However, even solutions that are worse can still be considered for acceptance, with a probability denoted as P. This probability is governed by the Boltzmann distribution, as elaborated in [16,19].

$$P = e^{-\frac{\Delta E}{kT_k}} \tag{4.17}$$

where ΔE denotes the variation of the objective function value, specifically $f_{opt,k+1} - f_{opt,k}$, where k represents Boltzmann constant and T_k stands for the current temperature value. Moving toward an inferior solution is allowed solely when P is greater than the value of a distributed random number r, drawn from the interval $[0,1]$. The most recent solution at temperature T_k serves as the preliminary solution for the subsequent step, where the temperature is gradually reduced (namely $T_{k+1} = \alpha T_k$, where $0 < \alpha < 1$).

This cyclic procedure concludes upon the attainment of a predetermined stopping criterion (specifically, the absolute change in objective function value falls below a fixed tolerance). On such occasions, no optimization is necessary, and the cooperative controllers solely compute the current value of the objective function. Should the controllers network detects a modification in the objective function (e.g. due to modifications of the grid conditions), the pursuit of a new optimal solution is warranted. During such instances, the temperature is set to its initial value, and the iterative search algorithm is re-initiated.

4.3.1 Example

Within this section, the effectiveness of the described solution framework in addressing the voltage regulation problem within the context of the IEEE 30-bus test system considered in Section 4.2.2 is evaluated.

The initial phase of the simulation studies involves evaluating the capability of the decentralized controllers network in solving the voltage regulation problem for the initial power system state. The obtained results are reported in Figures 4.10 and 4.11.

Particularly, Figure 4.10 reports the time evolution of the derivative state trajectories in the task of calculating the objective function (4.13). The corresponding regulating actions computed by the metaheuristic optimizers, employing the described SA-based minimization technique, are illustrated in Figure 4.11.

Upon analyzing these profiles, it is worth noting that decentralized metaheuristic optimizers converge to the actual objective function value within less than

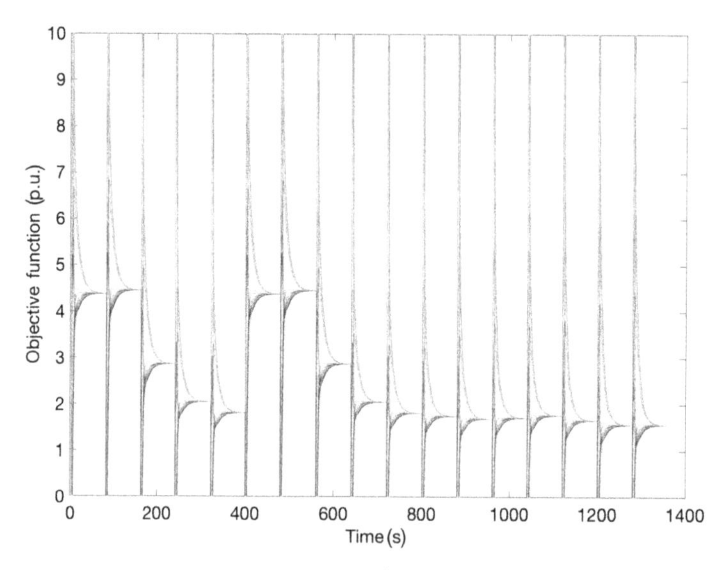

Figure 4.10 Derivative state trajectories ($\dot{\theta}_i$) self-synchronizing on the actual value of the objective function

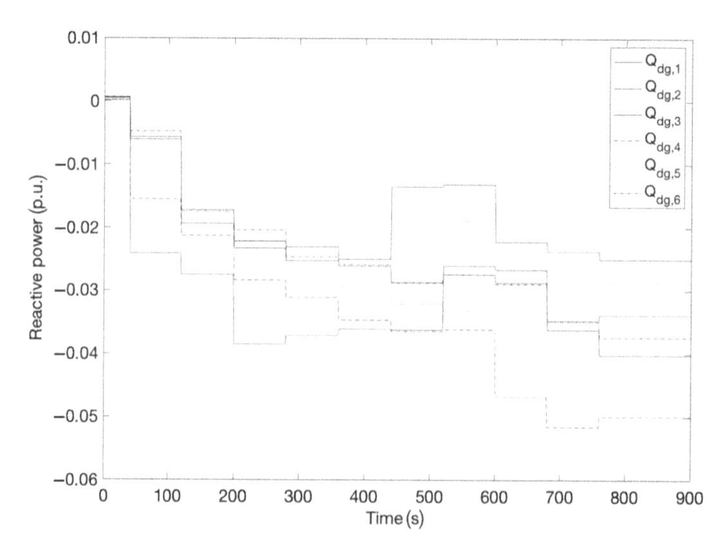

Figure 4.11 Voltage regulator set-points identified by the decentralized metaheuristic solution scheme

50 iterations. Moreover, they successfully identify the optimal voltage regulation strategy in fewer than 350 control iterations.

Also, in this case, the introduction of a nonlinear coupling function and a proper tuning of the coupling parameters is expected to significantly diminish the required convergence iterations.

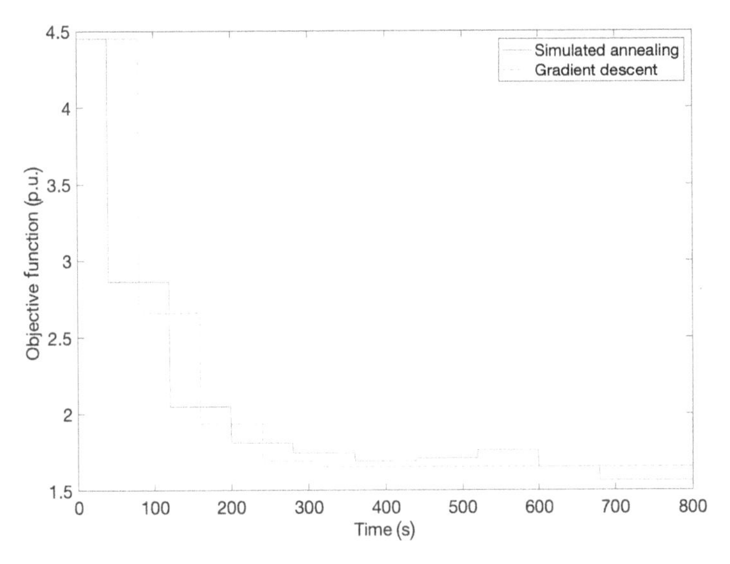

Figure 4.12 Evolution of the objective function

For the purpose of benchmarking the proposed approach, the optimal regulation strategy has also been determined by employing a conventional centralized solution scheme. It is important to emphasize that this comparative evaluation is not intended to showcase the superiority of the SA-based method over a centralized solution paradigm. Instead, it is assumed that the centralized approach provides an accurate solution to the voltage regulation problem. Hence, the comparison with the conventional solution scheme aims at assessing the degree of proximity between the achieved outcomes and those derived from the rigorous centralized approach, highlighting the advantages stemming from the integration of the SA-based search technique in the decentralized regulatory paradigm.

An overview of the results of this comparative analysis has been summarized in Figure 4.12.

The obtained results indicate that the solution derived from the SA-based local optimizers closely approximates the centralized solution obtained through the rigorous optimization algorithm. As expected, the centralized optimization approach exhibits superior convergence performance, but it necessitates a detailed power system model and a data fusion center responsible for collecting and processing all power system measurements.

In contrast, the decentralized solution schemes address the voltage regulation problem by embracing a fully decentralized and non-hierarchical approach. Remarkably, the actual objective function values are evaluated without dependency on a data fusion center, while the regulation strategies are determined by the decentralized optimizers that process both global and local variables.

In terms of a comparison with the outcomes of the distributed solution framework introduced in Section 4.2.2, it is evident that the SA-based approach demonstrates enhanced performance in terms of minimization efficacy. This is largely attributed to the efficacy of the SA-based minimization technique, which empowers probabilistic-based techniques to escape from local minima.

Consequently, a synergistic amalgamation of these two techniques could potentially elevate the overall performance of the decentralized regulation framework. Specifically, the gradient descent-based minimization approach might facilitate swift identification of a sub-optimal solution by the local optimizers, while the SA-based algorithm could contribute to augmenting the exploration of the solution space.

4.4 A decentralized fuzzy-based solution of the voltage regulation problem

Fuzzy logic-based solution schemes are recognized as a promising methodology for solving the voltage regulation problem in modern smart grids. Embracing this approach, we describe a decentralized and non-hierarchical voltage control framework that hinges on cooperative fuzzy agents. Drawing a parallel to self-organizing biological populations, voltage control is realized through local interactions among these fuzzy agents, according to the principles of distributed consensus theory. This empowers control agents to make proper decisions regarding the optimal injection of reactive power flow into the network, particularly in light of global network conditions.

In dealing with this matter, a conventional practice has been to employ a linear model of the power network, as observed in Ref. [20]. This model enables analysts to estimate the impact of voltage control actions by scrutinizing the sensitivity of the power flow solution.

Specifically, for an n-bus system, wherein buses 1 through N_D are designated as load buses, buses N_D through $n-1$ as generator buses, and bus n functions as the slack bus, the enhancement in voltage at bus i during the kth time step (denoted as $\Delta V_i = V_i(k+1) - V_i(k)$), stemming from adjustments to the control device at bus j (denoted as $\Delta U_i = U_i(k+1) - U_i(k)$), can be estimated as follows:

$$\Delta V_i = S_{ij} \Delta U_j \;\; \forall i \in [1, N_D] \; \forall j \in [N_D + 1, n-1] \tag{4.18}$$

where S_{ij} is the sensitivity coefficient of bus j on bus i.

Correct power system operation is contingent upon strictly adhering to various operational constraints. These encompass both the upper and lower thresholds for buses voltage magnitudes (specifically, $V_{min} \leq V_i \leq V_{max}$), the adjustable boundaries for the voltage regulator (namely, $\Delta U_{j,min} \leq \Delta U_j \leq \Delta U_{j,max}$), as well as their operational limitations (namely, $U_{j,min} \leq U_j \leq U_{j,max}$).

Employing the linear model (4.18) enables the estimation of diverse performance indicators that characterize the influence of each voltage control action on power system operation. As an illustration, the projected power loss stemming from the adjustment of the control device at bus j can be predicted using the following simplified equation [10]:

$$P_L^{(j)} = \sum_{k=1}^{N_L} g_k[(V_i+\Delta V_i)^2+(V_j+\Delta V_j)^2-2(V_i+\Delta V_i)(V_j+\Delta V_j)\cos(\delta_i-\delta_j)] \quad (4.19)$$

Here, N_L represents the number of lines, g_k denotes the conductance of the kth line connecting buses i and j, and δ_i as well as δ_j correspond to the voltages angles of bus i and j, respectively.

This model facilitates the identification of the set of feasible solutions (namely, voltage control strategies satisfying all network constraints) and the selection of the one that aligns with a preferred performance indicator (such as loss reduction).

In addressing this issue, the application of an approach grounded in fuzzy programming is here described.

4.4.1 Centralized solution

The effectiveness of fuzzy inference systems in managing imprecise data, coupled with their capacity to process qualitative information, establishes fuzzy logic as a viable solution framework for tackling the voltage control problem. Consequently, numerous research studies have been ignited, focusing on devising hybrid inference systems to enhance the efficacy of voltage and reactive power controllers [21,22].

Actually, the most prevalent fuzzy-based solution methodology for addressing the voltage control issue is organized into two computational phases [23]. In the initial phase, the principles of fuzzy sets are harnessed to identify a set of feasible solutions. Subsequently, in the second phase, a suitable selection criterion is employed to pinpoint the optimal control solution.

The concept revolves around introducing two fuzzy variables to characterize the degree of voltage violation for each bus experiencing a violation (namely, buses where the voltage magnitude deviates from the allowable limits) and the control capacity of each voltage control device. The corresponding membership functions are depicted in Figures 4.13 and 4.14, respectively.

The design of these membership functions has been based on insights from prior studies, as detailed in [23]. It is essential to highlight that we are interested in deploying a cutting-edge fuzzy paradigm within a network of dynamic and collaborative fuzzy agents, hence, we have employed the same parameters as the centralized fuzzy controller proposed in the literature, since the optimal tuning of the fuzzy control parameters is beyond the scope of this analysis.

In Figure 4.13, ΔV_i and $u_{\Delta V_i}$ are the voltage violation level and the corresponding membership function for the ith bus, respectively. In Figure 4.14, C_{ij} and $u_{C_{ij}}$ stand for the control capability of the device installed at bus j over bus i and its associated membership function. The control capability C_{ij} is defined in the following manner:

$$C_{ij} = S_{ij}M_j \quad (4.20)$$

where M_j represents the controlling margin of the voltage regulator installed at bus j.

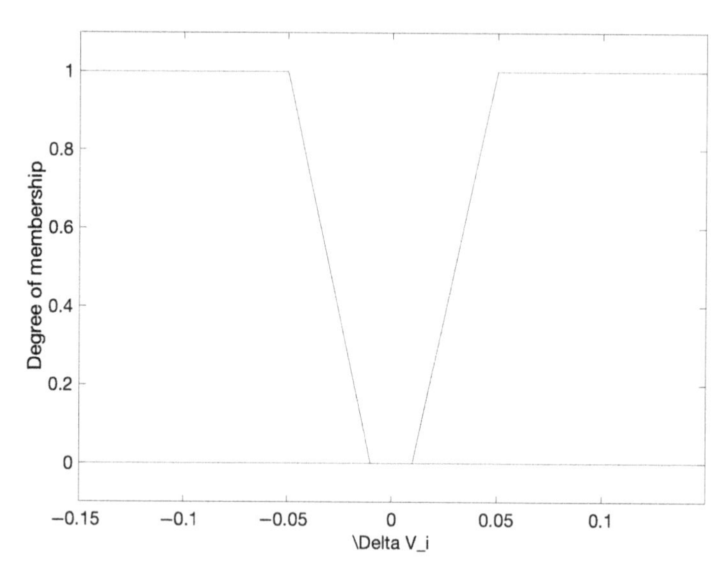

Figure 4.13 Membership functions for the control input ΔV_i

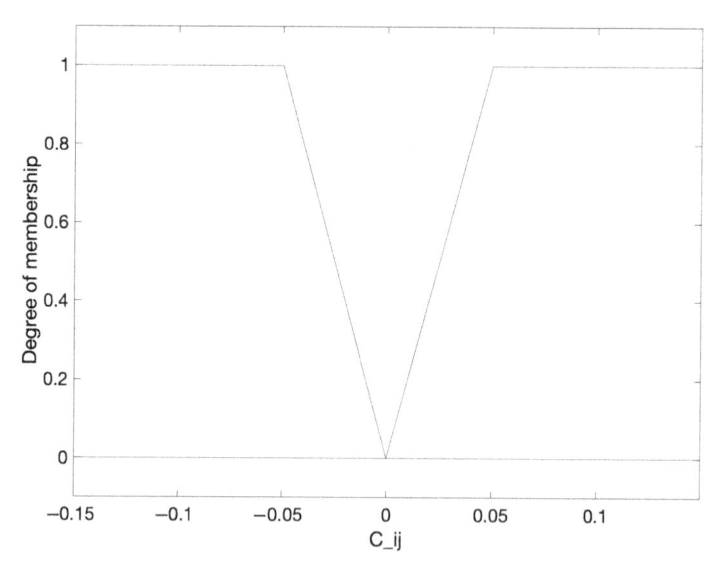

Figure 4.14 Membership functions for the control input C_{ij}

The feasible solution set assuring the grid voltage enhancement can be obtained by processing these fuzzy variables by employing the max–min operation:

$$R_{opt,j} = max\,min_i(u_{\Delta V_i}, u_{C_{ij}}) \;\; \forall i \in [1, N_D] \; \forall j \in [N_D, n-1] \tag{4.21}$$

where $R_{opt,j}$ describes the membership value of controlling ability for voltage regulators at bus j on the voltage magnitude of the bus i.

From these prospective control solutions, the subsequent step involves the application of appropriate selection criteria to discern the optimal control action. For this purpose, one potential method involves employing the maximum operator on the N_D components of $R_{opt,j}$ [24], namely

$$j^* = arg\ max_j\ R_{opt,j} \tag{4.22}$$

An alternative criterion entails choosing the regulation action with the least power loss among the feasible solutions [23]. For this purpose, the alteration in power loss associated with the modification of each control device should be evaluated using (4.19). Subsequently, the ultimate solution could be determined by

$$j^* = arg\ min\ \left(P_L^{(1)}, P_L^{(2)}, \ldots, P_L^{(n-N_D)}\right) \tag{4.23}$$

Here, $P_L^{(j)}$ represents the power loss that corresponds to an adjustment of the voltage regulator positioned at bus j by the quantity $R_{opt,j}$, while j^* designates the optimal device chosen for voltage control.

Upon identification of the optimal control solution, its impact on the power system undergoes evaluation through a power flow analysis. Should the voltage magnitudes at all buses adhere to network constraints, the control algorithm concludes. Conversely, if any buses still exhibit voltage irregularities, a new control iteration is initiated with the goal of removing these voltage anomalies.

This solution scheme is widely acknowledged as one of the most efficacious paradigms for fuzzy-based voltage regulation in power distribution systems [23,24]. It finds its roots in fuzzy programming theory and, in contrast to fuzzy control-based methods, it dispenses with the necessity of formulating fuzzy if-then rules and applying a defuzzification process.

4.4.2 Decentralizing the fuzzy operators by self-organizing sensors networks

Utilizing the described consensus algorithms empowers the control agents to determine the optimal set-points of the voltage regulators, aligned with the fuzzy-based solution algorithm expounded in Section 4.4.1.

For this purpose, several issues should be solved, such as

- The calculation of average, minimum, and maximum voltage magnitudes at load buses is pivotal for agents to trigger the regulation adjustment process.
- Every agent needs to be aware of the allowable control range for each voltage regulation device, which enable the computation and fuzzification of their respective control capabilities.
- The controllers should implement the desired selection criteria by evaluating R_{j^*}.

To tackle the first concern, each agent must acquire the local bus voltage magnitude. This measurement is organized according to the following observation vector [19]:

$$\omega_i = [V_i, V_i, V_i] \tag{4.24}$$

In particular, the first component of the observation vector is used to solve an average consensus problem, while the other two evolve according to a maximum/minimum consensus protocol.

In particular, as far as the consensus protocol adopted for self-synchronization on the maximum value is concerned, it is based on the following coupling strategy:

$$\begin{cases} \theta_i(k+1) = \max(\theta_i(k), u_i(k)) \\ u_i(k) = \max_{j \in \mathcal{N}_i} \theta_i(k) \end{cases} \tag{4.25}$$

This strategy allows all of the dynamic oscillators to quickly synchronize on the observation value of the max leader, which is the oscillator characterized by the highest observed value [25]. The same coupling strategy can be deployed for self-synchronization on the minimum observed value.

Consequently, after reaching the consensus, the state of all the dynamic oscillators synchronize to the following values:

$$\dot{\theta} = \left[\frac{\sum_{i=1}^{n} V_i}{n}, \max_i V_i, \min_i V_i \right] \tag{4.26}$$

To solve the second issue, each control agent should share its state with all the controllers network. This can be obtained by solving an average consensus problem using the following observation vector:

$$\begin{cases} \omega_i = [0, ..., 0, ..., 0] \ \forall i \in [1, N_D] \\ \omega_i = [0, ..., M_i, ..., 0] \ \forall i \in [N_D + 1, n] \end{cases} \tag{4.27}$$

In this case, when the dynamic oscillators synchronize to the following state:

$$\dot{\theta} = [nM_{N_D+1}, nM_{N_D+2}, ..., nM_n] \tag{4.28}$$

Finally, to solve the third issue, the observation vector should be defined according to one of the following equations, which depend on the particular selection criteria that should be implemented:

$$\begin{cases} \omega_i = R_{opt,i} \ \forall i \in [1, N_D] \\ \omega_i = P_L^i \ \forall i \in [1, N_D] \end{cases} \tag{4.29}$$

The optimal solution to the voltage regulation problem is then obtained by solving a maximum and minimum consensus problem, respectively.

4.4.2.1 Example

In this section, the core findings resulting from the application of the decentralized fuzzy-based approach for voltage regulation within the IEEE 30-bus test system described in Section 4.2.2 are detailed and examined. The primary control objective aims at achieving coordination among the six voltage regulators to uphold the voltage magnitude within an acceptable range (specifically, within 5% of the nameplate value) for each load bus.

In the initial phase of our experiments, our focus is on evaluating the effectiveness of the decentralized fuzzy controllers in accurately assessing the actual state of the power system.

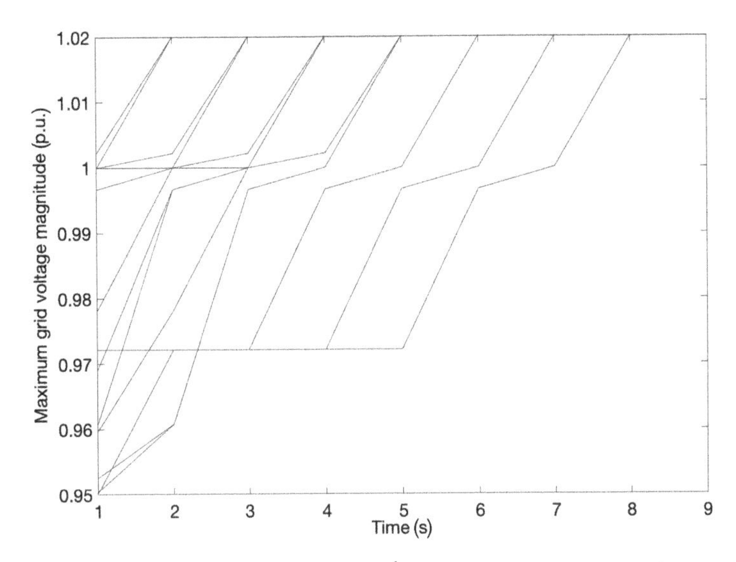

Figure 4.15 Derivative state trajectories ($\dot{\theta}_i$) self-synchronizing on the maximum value of the grid voltage magnitude

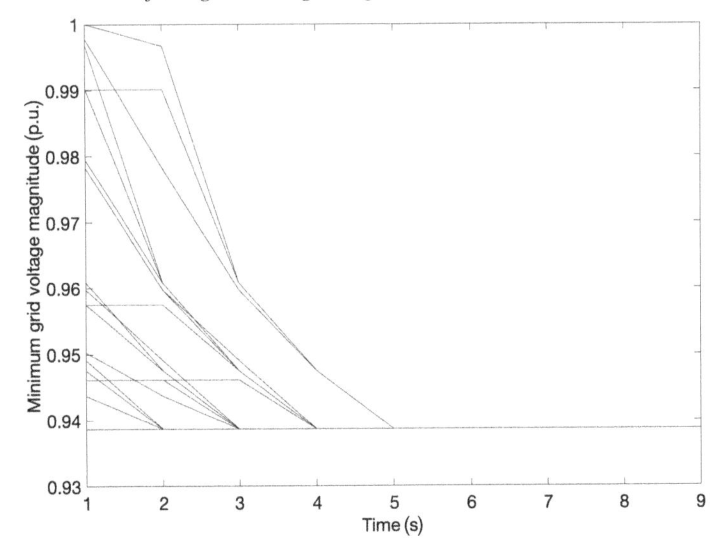

Figure 4.16 Derivative state trajectories ($\dot{\theta}_i$) self-synchronizing on the minimum value of the grid voltage magnitude

As expected, the implementation of the distributed consensus protocols brings about the convergence of all dynamic oscillators (initially assigned random states) toward the real values of grid variables (namely, max and min grid voltage), as reported in Figures 4.15 and 4.16, respectively.

With regard to the synchronization time, it is projected to be within a few seconds for wired communication setups, which appears to align well with the time constraints intrinsic to the voltage control process in smart grids.

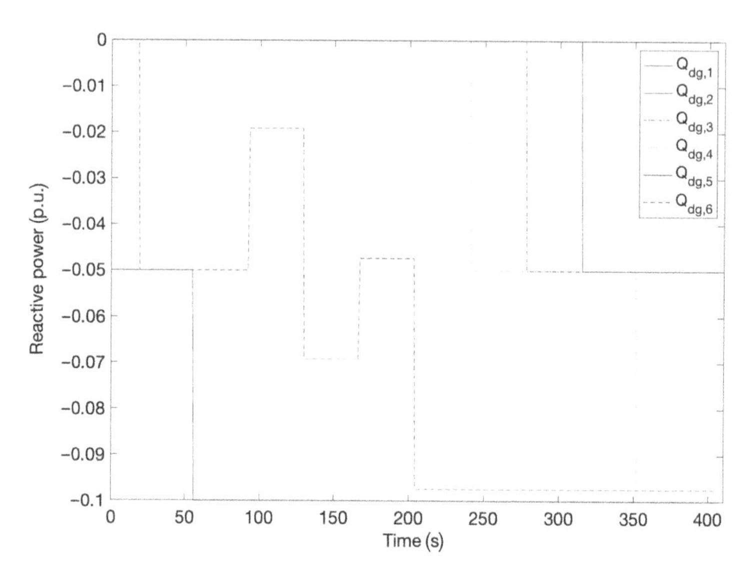

Figure 4.17 Voltage controllers output

Another important consideration while assessing the performance of the fuzzy-based solution approach pertains to the quantity of messages exchanged between agents. In particular, the application of the decentralized framework mandates each agent to transmit a message to its neighboring agents during each iteration. Consequently, the exchange of messages among agents to attain consensus on a general function is likely to be higher compared to a centralized approach. This drawback is well-known within the context of decentralized consensus protocols and, generally speaking, with any kind of distributed control architectures. In our viewpoint, scientific methodologies should be formulated for the task of minimizing the volume of exchanged messages in decentralized control architectures.

To gauge the effectiveness of the coordination among voltage controllers, the network of agents was activated in solving the voltage control problem removing the critical voltage anomalies across the various load buses. The outcomes obtained from this testing are summarized in Figures 4.17—4.19.

Particularly, Figure 4.17 illustrates the time profiles of the voltage regulator set-points, as identified by the distributed fuzzy controllers.

Upon analysis of these graphs, it is noteworthy that the voltage regulation strategy determined by the distributed fuzzy controllers necessitates just eight control iterations to significantly enhance the bus voltage profile. Additionally, it is crucial to emphasize that this outcome arises from a fully decentralized control architecture, as all the variables required for implementing the fuzzy-based solution algorithm have been assessed by the agents without reliance on a fusion center. This affirmation is corroborated by Figures 4.18 and 4.19, which present the evolution of agent states during control margin evaluation and optimal control solution selection tasks.

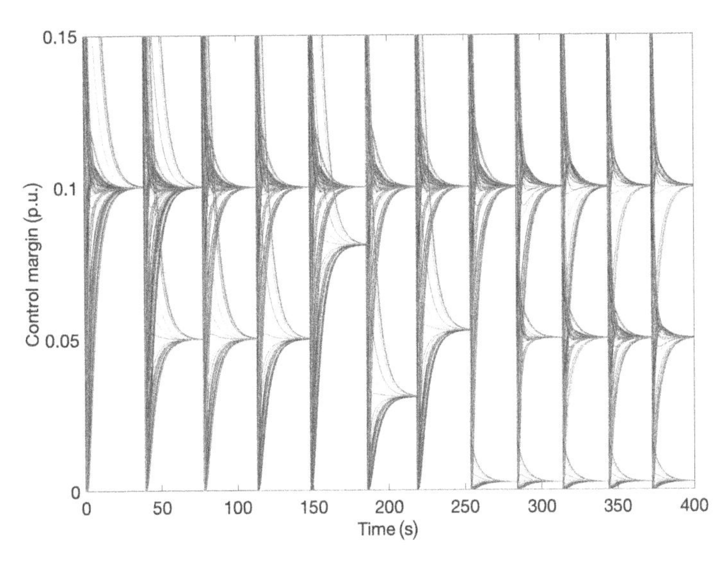

Figure 4.18 Evolution of the agent states in the task of estimating the control margin

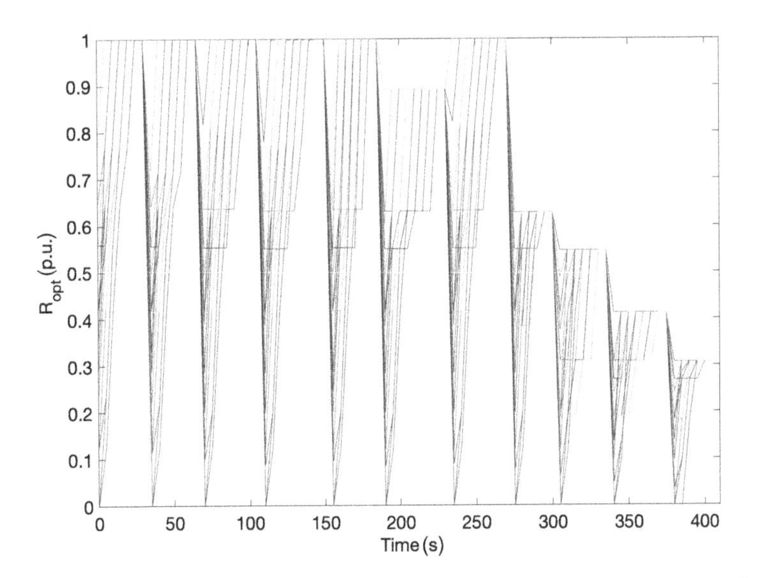

Figure 4.19 Evolution of the agent states in the task of selecting the optimal control action

Regarding algorithm complexity, it is essential to recognize that it is predominantly linked to solving a system of interdependent first-order differential equations, with their convergence governed by the eigenvalues of the adjacency matrix. In initial experimental studies, response times on the order of tenths of seconds were observed.

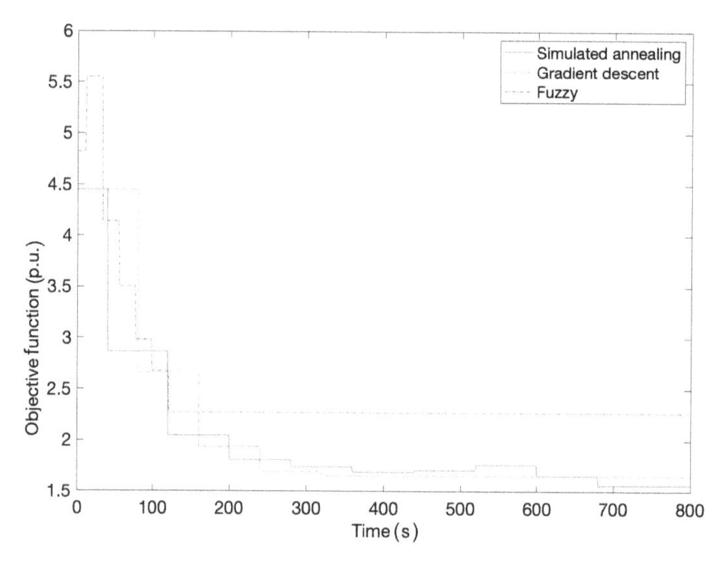

Figure 4.20 Results comparison

These response times were gauged using a sniffer node that captured and processed all agent states.

Lastly, for the purpose of establishing a benchmark against the described approach, the optimal regulation strategy has also been determined using a cutting-edge solution algorithm based on the Interior Point Method. Employing detailed mathematical model of the analyzed power network, this rigorous centralized solution methodology identifies, for each network state, the optimal set-points of the voltage regulators by solving a constrained optimization problem. In addressing this challenge, its aim is to minimize a scalar cost function defined by a linear combination of control objectives, namely, voltage profile optimization and minimization of active power losses. The obtained results are reported in Figure 4.20.

4.4.3 Decentralized fuzzy-rule-based controllers

This decentralized control algorithm enables each voltage controller to regulate the reactive power flows injected by the distributed generators into the electrical grid. To estimate the current grid state, the algorithm first estimates the global variables that describe the Smart Grid actual operation, using the bio-inspired paradigm mentioned earlier. These variables are then processed to classify the current grid state based on the mean grid voltage magnitude, using the following grid state classification [26]:

1. "Under-voltage": This grid operation state is marked by a significant power demand, leading to high-power line loading, which results in increased power losses and low bus voltage magnitudes.

2. “Over-voltage”: This grid operation state is marked by low power demand, leading to low power line loading, which results in decreased power losses and a high bus voltage magnitudes.
3. “Acceptable-voltage”: When the electrical grid operates under these conditions, the grid is considered to be at its nominal operating point, and both the power losses and the mean grid voltage magnitude remain within acceptable limits.

After evaluating the current state of the smart grid, proper voltage control strategies for each node are determined by processing both global and local variables. Subsequently, these strategies are put into practice by following specific control rules:

- If the power system state is classified as “Under-voltage” (respectively, “Over-voltage”), then the main goal is to raise (respectively, lower) the average system voltage magnitude V_M by raising (respectively, lowering) the reactive power injected by the distributed generators into the grid. For this purpose, in order to obtain a uniform rise of the power system voltage magnitude, the ith agent compares the measured bus voltage magnitude (V_i) with the average grid voltage magnitude (V_M) and:
  - if $V_i < V_M$ it raises (respectively, lowers) the injected reactive power of the controlled generator $Q_{dg,i}$ according to the following control law:
  $$\Delta Q_{dg,i} = \alpha_i(V_i)(V_M - V_i) \text{ respectively } \Delta Q_{dg,i} = \frac{\alpha_i(V_i)}{V_i - V_M} \tag{4.30}$$
  - Otherwise, it raises (respectively, lowers) injected reactive power of the controlled generator $Q_{dg,i}$ according to the following control law:
  $$\Delta Q_{dg,i} = \frac{\alpha_i(V_i)}{V_i - V_M} \text{ respectively } \Delta Q_{dg,i} = \alpha_i(V_i)(V_M - V_i) \tag{4.31}$$
- If the power system state is classified as “Acceptable-voltage,” then the main goal is to enhance the grid voltage profile by minimizing the bus voltage magnitude deviation. For this purpose, the ith agent compares the measured voltage magnitude (V_i) with the mean power system voltage magnitude (V_M) and:
  - if $V_i < V_M$ it raises the reactive power $Q_{dg,i}$ according to the following control law:
  $$\Delta Q_{dg,i} = \alpha_i(V_i)(V_M - V_i) \tag{4.32}$$
  - Otherwise, it lowers the reactive power $Q_{dg,i}$ according to the following control law:
  $$\Delta Q_{dg,i} = \alpha_i(V_i)(V_M - V_i) \tag{4.33}$$

  where $\alpha_i(V_i)$ is a monotonically increasing function.

One way to implement these user-supplied human language rules is by using a fuzzy inference system. This system converts the linguistic rules into mathematical equivalents, making it easier for the system designer to accurately represent the behavior of the voltage control system in realistic operation scenarios [26]. Fuzzy logic has additional benefits such as simplicity and flexibility. It can handle problems with imprecise and incomplete data and can model nonlinear functions of any complexity. These advantages make it a suitable candidate for integration in distributed voltage controllers.

4.4.3.1 Example

In this section, we will discuss the use of the described fuzzy rules-based control strategy for voltage regulation in the IEEE 30-bus test system. Our experiments considered six dispatchable distributed generation sources, and each variable was regulated by a voltage controller. Hence, six control variables have been defined:

$$y = [Q_{dg,1}, Q_{dg,2}, Q_{dg,3}, Q_{dg,4}, Q_{dg,5}, Q_{dg,6}] \tag{4.34}$$

To evaluate the global variables that characterize the power grid operation, we deployed a sensor network consisting of 30 cooperative sensors distributed throughout the power system, which measure the bus voltage magnitude, and the generated active and reactive power. Hence, the corresponding observation vector is defined as

$$\omega_i = (V_i, 30P_i, |V_i - V_i^*|) \tag{4.35}$$

It allows the dynamic systems to synchronize to the following values:

$$\omega^* = \left(\sum_{i=1}^{3} 0 \frac{V_i}{30}, \sum_{i=1}^{3} 0 P_i, \sum_{i=1}^{3} 0 \frac{|V_i - V_i^*|}{30} \right) \tag{4.36}$$

The coupling coefficients aij, which describe the coupling of sensors, were obtained using the same adjacency matrix characterizing the electrical network.

The simulation results showed that the built-in dynamical systems synchronized to the values representing mean grid voltage magnitude, total power losses, and average voltage magnitude deviation in about 50 iterations. The overall synchronization time was of the order of 3 s for wired communications, which is suitable for the voltage regulation process in smart grids.

Simulation analyses have been carried out to evaluate the regulation performance of the electrical grid under different load patterns. The evolution of regulating variables and their impact on bus voltage profiles have been reported in Figures 4.21 and 4.22, respectively. The distributed fuzzy controllers have been used to identify voltage regulation strategies for a critical grid state. It is worth mentioning that after 30 iterations, the voltage regulation strategy identified by the controllers has led to a significant improvement in the bus voltage profile. This observation is supported by Figures 4.23 and 4.24, which report the corresponding evolution of the mean grid voltage magnitude and the active power losses, respectively. These variables are key indicators of the power grid overall performance, and thus, should be taken into account when devising an effective solution to the voltage regulation problem. The results demonstrate that the regulation strategy has a positive impact on the entire grid, as evidenced by the tendency of mean grid voltage magnitude towards its nameplate value (1 p.u.) and a reduction in active power losses. Notably, the described fuzzy-rule-based architecture adopts a fully decentralized regulating approach, where global variables characterizing the power grid operation have been assessed without a fusion center. The regulation strategies have been identified using distributed fuzzy inference systems that process both global and local variables. This decentralized approach renders the architecture highly scalable, self-organizing, and distributed.

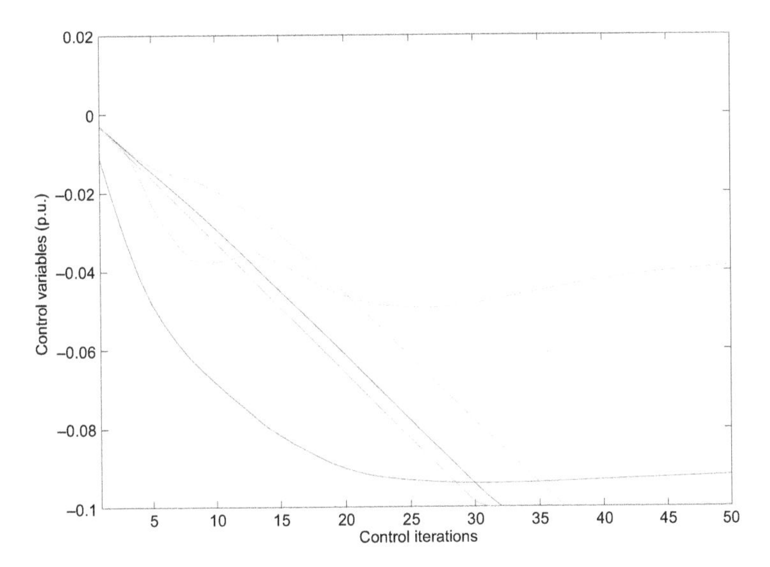

Figure 4.21 Evolution of controlled variables

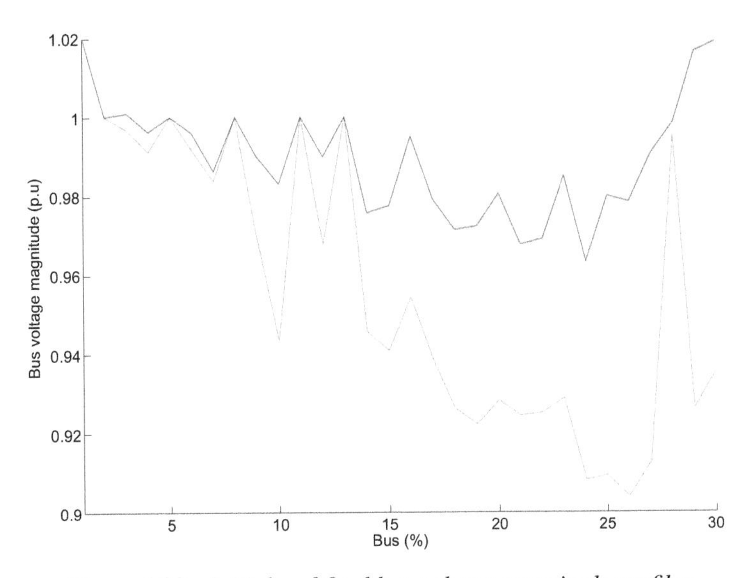

Figure 4.22 Initial and final bus voltage magnitude profiles

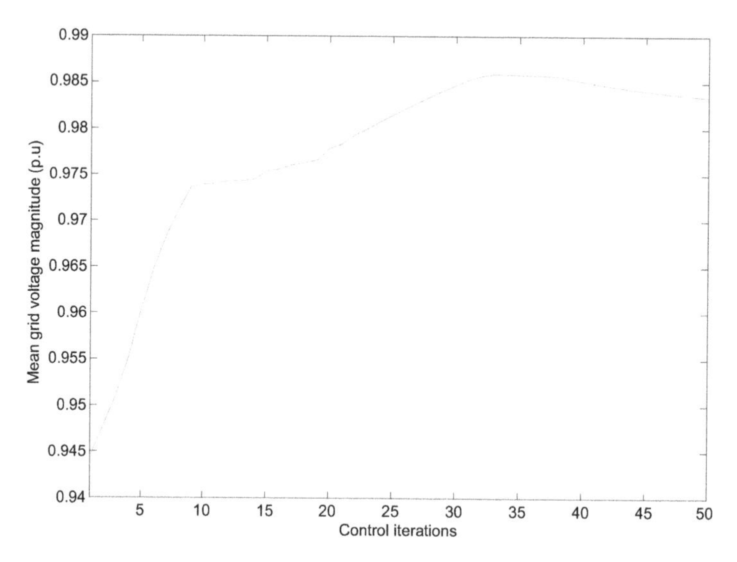

Figure 4.23 Evolution of mean grid voltage magnitude

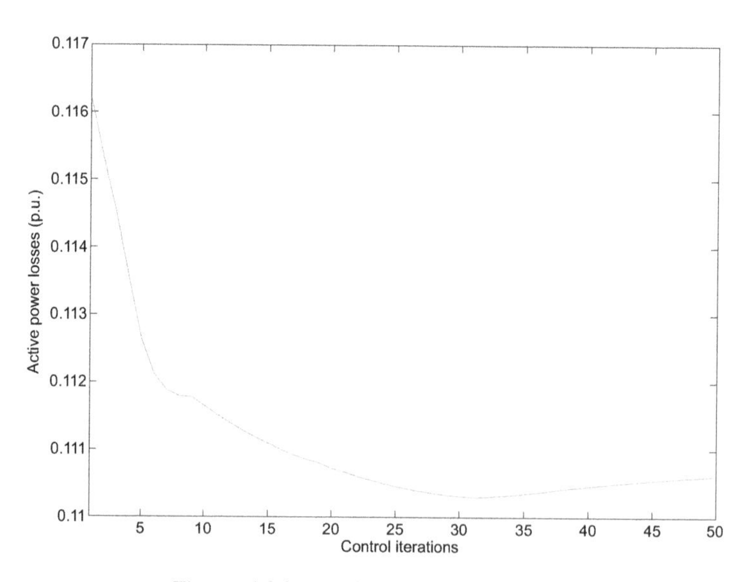

Figure 4.24 Evolution of power losses

References

[1] Vaccaro A, Velotto G, and Zobaa AF. A decentralized and cooperative architecture for optimal voltage regulation in smart grids. *IEEE Transactions on Industrial Electronics*. 2011;58(10):4593–4602.
[2] IEEE Smart Grid Vision for Computing: 2030 and Beyond Reference Model. 2016. p. 1–18.
[3] Suhaimy N, Radzi NAM, Ahmad WSHMW, *et al.* Current and future communication solutions for smart grids: a review. *IEEE Access*. 2022;10: 43639–43668.
[4] Sun X, Li T, and Xing L. Research on the influence of microgrid to distribution network protection and improvement measures. In: *2020 IEEE Sustainable Power and Energy Conference (iSPEC)*; 2020. p. 2243–2248.
[5] Jain H, Palmintier B, Krad I, *et al.* Studying the impact of distributed solar PV on power systems using integrated transmission and distribution models. In: *2018 IEEE/PES Transmission and Distribution Conference and Exposition (T&D)*; 2018. p. 1–5.
[6] Palaniappan R, Funke A, Hilbrich D, *et al.* A robust and resilient voltage control strategy for smart grids using distributed measurements. In: *2019 IEEE Innovative Smart Grid Technologies – Asia (ISGT Asia)*; 2019. p. 2411–2415.
[7] Yang J, Tushar W, Saha TK, *et al.* Prosumer-driven voltage regulation via coordinated real and reactive power control. *IEEE Transactions on Smart Grid*. 2022;13(2):1441–1452.
[8] Chai Y, Guo L, Wang C, *et al.* Hierarchical distributed voltage optimization method for HV and MV distribution networks. *IEEE Transactions on Smart Grid*. 2020;11(2):968–980.
[9] Fu A, Cvetković M, and Palensky P. Distributed cooperation for voltage regulation in future distribution networks. *IEEE Transactions on Smart Grid*. 2022;13(6):4483–4493.
[10] Hareesh Kumar Reddy M and Vignesh V. Graph theoretic approach for decentralized control architecture of cyber physical smart grid. In: *2023 11th Workshop on Modelling and Simulation of Cyber-Physical Energy Systems (MSCPES)*; 2023. p. 1–6.
[11] Boiarkin V, Asif W, and Rajarajan M. Decentralized demand response power management system for smart grids. In: *2020 IEEE 8th International Conference on Smart Energy Grid Engineering (SEGE)*; 2020. p. 70–74.
[12] Werner S, Walter T, Wiezorek C, *et al.* Smart decentralised energy management. In: *CIRED 2020 Berlin Workshop (CIRED 2020)*. vol. 2020; 2020. p. 345–348.
[13] Vaccaro A, Loia V, Formato G, *et al.* A self-organizing architecture for decentralized smart microgrids synchronization, control, and monitoring. *IEEE Transactions on Industrial Informatics*. 2015;11(1):289–298.
[14] Loia V and Vaccaro A. A decentralized architecture for voltage regulation in Smart Grids. In: *2011 IEEE International Symposium on Industrial Electronics*; 2011. p. 1679–1684.

[15] Fliege J and Fux Svaiter B. Steepest descent methods for multicriteria optimization. *Mathematical Methods of Operations Research*. 2000;51(1): 479–494.

[16] Galdi V, Vaccaro A, and Villacci D. Voltage regulation in MV networks with dispersed generations by a neural-based multiobjective methodology. *Electric Power Systems Research*. 2008;78(5):785–793.

[17] Gerard M, Schutter BD, and Verhaegen M. A hybrid steepest descent method for constrained convex optimization. *Automatica*. 2009;45(2):525–531.

[18] Sousa T, Morais H, Castro R, *et al.* A new heuristic providing an effective initial solution for a simulated annealing approach to energy resource scheduling in smart grids. In: *2014 IEEE Symposium on Computational Intelligence Applications in Smart Grid (CIASG)*; 2014. p. 1–8.

[19] Vaccaro A and Zobaa AF. Voltage regulation in active networks by distributed and cooperative meta-heuristic optimizers. *Electric Power Systems Research*. 2013;99:9–17.

[20] Tomsovic K. A fuzzy linear programming approach to the reactive power/voltage control problem. *IEEE Transactions on Power Systems*. 1992;7(1): 287–293.

[21] Zhang W and Liu Y. Multi-objective reactive power and voltage control based on fuzzy optimization strategy and fuzzy adaptive particle swarm. *International Journal of Electrical Power & Energy Systems*. 2008;30(9):525–532.

[22] Miranda V, Moreira A, and Pereira J. An improved fuzzy inference system for voltage/VAR control. *IEEE Transactions on Power Systems*. 2007;22(4):2013–2020.

[23] Rahideh A, Gitizadeh M, and Rahideh A. Fuzzy logic in real time voltage/reactive power control in FARS regional electric network. *Electric Power Systems Research*. 2006;76(11):996–1002.

[24] Su CT and Lin CT. Fuzzy-based voltage/reactive power scheduling for voltage security improvement and loss reduction. *IEEE Transactions on Power Delivery*. 2001;16(2):319–323.

[25] Olfati-Saber R and Murray RM. Consensus problems in networks of agents with switching topology and time-delays. *IEEE Transactions on Automatic Control*. 2004;49(9):1520–1533.

[26] Loia V and Vaccaro A. A decentralized architecture for voltage regulation in smart grids. In: *2011 IEEE International Symposium on Industrial Electronics*; 2011. p. 1679–1684.

Chapter 5
Decentralized economic dispatch of distributed generators

The solution of economic dispatch (ED) problems holds a central and extensively applied position in the context of smart grids analysis. The primary objective of ED is to efficiently manage the generation unit production, ensuring the system load demand is met at the least operational cost, while adhering to generators and grid constraints. The overall problem can be cast as a constrained nonlinear optimization problem [1], which can be solved by using numerous solution techniques spanning iterative numerical methods grounded on the Lagrange multipliers theory [2], dynamic programming [3], evolutionary algorithms [4], and heuristic approaches [5].

While these strategies provide valuable insights into the role of modern optimization techniques within ED analysis, their application necessitates the establishment of a data fusion center responsible for collecting and processing all the needed power system measurements. An ongoing debate within the smart grids research community delves into the appropriateness of this hierarchical control paradigm, particularly in light of the radical modification affecting modern power systems.

Specifically, the proliferation of smart grids technologies is expected to dramatically increase data exchange rates, straining centralized control structures to rapid saturation. Consequently, the data collected from distributed grid sensors may not offer timely guidance to system operators, despite the deployment of advanced data-to-information conversion tools. Moreover, the communication networks connecting the sensors to the power system control centers may be frequently saturated, requiring proper redundancies in order to enhance their resilience to external cyber-attacks.

In this context, hierarchical control paradigms might not be suitable to cope with the increasing grid complexity and the pervasive presence of distributed generators inherent in modern smart grids [6]. Hence, the development of distributed multi-agent optimization paradigms has gained prominence, given their successful application in enhancing the operational efficiency of complex systems through decentralized and collaborative agent networks [7].

With this vision in mind, various studies highlight the role of multi-agent systems in addressing smart grid operation problems, including optimal power flow studies, ED [6], load restoration [8], and voltage regulation. These analyses emphasize the potential for distributed control architectures to enhance power systems efficacy by reducing dependencies and bolstering operational resilience following disturbances or equipment loss [8]. Furthermore, multi-agent systems-based architectures exhibit stability and self-healing features.

From our perspective, these attributes hold substantial promise in the context of ED analysis within smart grids, not solely for parallelizing global-scale solution algorithms, but also for distributing intelligence across distributed power generators. Moreover, we contend that the search for rigorous tools and scientific methodologies to define a fully decentralized and self-organizing control architecture remains in its nascent stages and requires further exploration.

To address this need, this chapter describes the integration of self-organizing dynamic agents equipped with distributed consensus protocols to tackle ED analysis. Specifically, as demonstrated in [9], we show that, under certain assumptions, solving the ED problem can be achieved by computing weighted averages of grid variables. To perform these computations, we describe the deployment of a dynamic agent network that solves distributed consensus problems.

Each agent in this network communicates only with a limited number of neighbors and possesses an oscillator initialized with local information. The oscillators of neighboring agents are interlinked via the bio-inspired protocols described in Chapter 2. We illustrate that the fundamental operations needed to solve the ED problem can be readily computed by agents following a completely decentralized and non-hierarchical paradigm.

We anticipate that this bio-inspired solution approach offers several advantages over conventional client–server paradigms, including reduced network bandwidth requirements, decreased computation time, and ease of extension and reconfiguration. To validate the efficacy of this framework, we present and discuss simulation results obtained from several power system test networks.

5.1 ED in smart grids

Distributed generators scheduling can be formalized as an optimization problem that encompasses an objective function, a series of design variables, and constraints. The specific set of design variables and constraints of this optimization problem is dependent on the components that need to be modeled. In the case of smart grids, these elements include multiple sources of flexibility that can be leveraged for optimal scheduling programs. These sources include household appliances such as dishwashers and washing machines, which can be scheduled to shift their activation times based on input data. Heating, ventilation, and air conditioning systems are also a flexible source as they can be controlled through energy management systems and their consumption is highly elastic. Distributed generation and energy storage systems are additional sources of flexibility that can be utilized for demand response programs. Furthermore, electric vehicles equipped with vehicle-to-grid (V2G) functions are expected to become a significant source of flexibility in the residential sector. The charging and discharging profiles of plug-in hybrid electric vehicles (PHEV) and electric vehicles can be controlled to support smart grid flexibility.

The objective function is typically aimed at minimizing electricity costs by means of optimal scheduling or maximizing social welfare when the utility is a part of the optimization problem. Another objective function is the reduction of load peak by minimizing the aggregated power consumption, which is primarily a distributed system operator management issue. These objectives can also be combined.

The mathematical formalization of objective functions and constraints can vary based on the chosen model for optimal scheduling. Convex optimization is the basic model that requires linear or convex objective functions and constraints. Traditional convex optimization techniques can be used to solve these models effectively. However, to obtain a more accurate model that allows for modeling on/off binary variables, the problem can be cast as a MILP [10]. This is often the case in demand response studies, where utilities assume the role of influencing customer behavior through price signals, assuming that customers have no role in setting prices. But, when this assumption is not valid, game theoretic models must be employed to consider the interaction between customers and utilities [11].

Game theoretic-based scheduling studies consist of a set of players maximizing their payoff by choosing an optimal scheduling strategy. The players involved are typically an utility and a set of customers who interact reciprocally until an optimal price is set. Stackelberg games can be used to model this interaction, where the utility is the leader that takes the first decision, and the customers are the followers that react to the leader's decision. The optimal point for both the utility and the customers is the Nash equilibrium, which is the convergence point.

In optimal scheduling, there are often a large number of time-correlated decision variables due to the presence of time-varying parameters, especially when real-time pricing is considered [12]. Dynamic programming can be used to handle these cases. However, in highly nonlinear models, heuristic methods are often the best choice to find the optimal solution [13]. Examples of such methods include genetic algorithms and simulated annealing. Although these methods have lower computational complexities and do not require perfect information as input, they may not guarantee the calculation of a global optimum.

When predictions are involved in the scheduling process, uncertainty from variable renewable energy sources (VRES), load, and prices can be modeled through robust or stochastic optimization programs. Stochastic methods use probability distribution functions associated with uncertain variables to find the optimal scheduling over a set of possible scenarios [14,15]. On the other hand, robust optimization provides more conservative solutions and is less computationally burdensome. These methods are more suitable for commercial and industrial users who may incur large economic losses due to forecasting errors.

5.2 Problem formulation

The core objective of ED analysis is to evaluate the most efficient production level for a set of dispatchable power generators. This production level should align with the system load demand while minimizing the generation costs and ensuring correct power system operation. The overall problem can be mathematically expressed through the subsequent constrained nonlinear optimization problem: [3]

$$\begin{aligned} \min_{x} \quad & F_T(x) \\ \text{subject to} \quad & g_i(x) \leq 0, \ \forall i = [1, m] \end{aligned} \tag{5.1}$$

where $g_i \ \forall i = [1, m]$ are the constraint functions describing the correct power system operation, and F_T represent the cost function, which describes the generation cost, and can be expressed as:

$$F_T = \sum_{i=1}^{n_g} C_i(P_i) \tag{5.2}$$

where n_g indicates the total number of the dispatchable generators, while P_i and C_i are the generated active power and the production cost function of the ith generator, respectively.

The production cost functions C_i can be expressed by polynomial functions [4]:

$$C_i(P_i) = \alpha_i + \beta_i P_i + \gamma_i P_i^2 \tag{5.3}$$

where α_i, β_i, and γ_i represent the approximating coefficients of the ith cost function.

Regarding the system constraints, they encompass both equality conditions, such as the active power balance constraint, and inequality constraints, such as the limitations on generator capacity.

In its most basic formulation, the ED problem neglects the effects of network losses and generator limits. This over-simplification empowers analysts to approach ED analysis by solving the subsequent programming problem:

$$\begin{aligned} \min_{P_1, \ldots, P_{n_g}} \quad & \sum_{i=1}^{n_g} C_i(P_i) \\ \text{subject to} \quad & \sum_{i=1}^{n} P_i = P_D \end{aligned} \tag{5.4}$$

where n ($n \geq n_g$) is the total number of committed generators and P_D represents the system demand.

To encompass both the constraints posed by generator limits and network losses, it becomes essential to introduce a more realistic formalization of the problem [9]

$$\begin{aligned} \min_{P_1, \ldots, P_{n_g}} \quad & \sum_{i=1}^{n_g} C_i(P_i) \\ \text{subject to} \quad & \sum_{i=1}^{n} P_i = P_D + P_L(P_1, \ldots, P_{n_g}) \\ & P_{i,\min} \leq P_i \leq P_{i,\max}, \ \forall i \in [1, n_g] \end{aligned} \tag{5.5}$$

where $P_{i,\min}$ and $P_{i,\max}$ are the minimum and maximum values of the active power that can be generated by the ith generator, and P_L represents the power system losses, which can be determined by considering the power flow equations, or approximated by using simplified expressions.

5.3 A centralized solution of the ED problem

The solution algorithm considered as a benchmark in our analysis is grounded on the principles of the Lagrange multiplier theory. Within the framework of this algorithm, the ED problem described in (5.4) can be addressed by identifying the stationary points of the following Lagrangian function [2]:

$$\mathscr{L} = \sum_{i=1}^{n_g} C_i(P_i) + \lambda(P_D - \sum_{i=1}^{n} P_i) \tag{5.6}$$

To solve this problem, it is possible to deploy an iterative solution approach aimed at enhancing the current solution denoted as $x = [P_1, ...P_{n_g}, \lambda]$ by solving the subsequent set of linear equations:

$$H_x[\Delta P_1, ..., \Delta P_{n_g} \Delta \Lambda] = \nabla \mathscr{L}|_x \tag{5.7}$$

where $\mathscr{L}|_x$ and H_x are the gradient vector and the Hessian matrix evaluated at the current solution x, respectively.

This solution scheme can be generalized in order to solve the ED problem by considering the effects of the system losses and the generation capacity constraints. In particular, modeling the system losses requires the integration of the power flow equations into the set of equality constraints. While this strategy provides valuable insights into power system operation, its practical application might be hindered by the substantial computational resources it demands. Particularly in scenarios involving distributed and cooperative computing entities, the resource requirements could pose challenges. To overcome this, we delve into simplified methodologies aimed at computing active power losses. Specifically, we explore the utilization of the Kron's loss formula formalized in (5.8), and the simplified expression formalized in (5.9) as part of our investigation. Notably, both approaches assume constant values for the loss coefficients B_{ij} and B_i.

$$P_L = \sum_{i=1}^{n_g} P_i \sum_{j=1}^{n_g} B_{ij} P_j + \sum_{i=1}^{n_g} B_{0i} P_i + B_{00} \tag{5.8}$$

$$P_L = \sum_{i=1}^{n_g} B_i P_i^2 \tag{5.9}$$

Moreover, the generation capacity constraints can be modeled by considering a proper number of inequality constraints, each one requiring the integration of additional Lagrange multipliers (namely $\mu_{i,\min}$ and $\mu_{i,\max}$) in the Lagrange function, which is now expressed as

$$\mathscr{L} = \sum_{i=1}^{n_g} C_i(P_i) + \lambda(P_D + P_L - \sum_{i=1}^{n} P_i) + \sum_{i=1}^{n_g} \mu_{i,\max}(P_i - P_{i,\max}) + \sum_{i=1}^{n_g} \mu_{i,\min}(P_{i,\min} - P_i) \tag{5.10}$$

The necessary conditions guaranteeing the solution to this problem, which are referred as the Karush–Kuhn–Tucker conditions, can be expressed as

$$\begin{cases} \frac{\partial \mathscr{L}}{\partial P_i} = 0 \\ \frac{\partial \mathscr{L}}{\partial \lambda} = 0 \\ \frac{\partial \mathscr{L}}{\partial \mu_{i,\max}} = P_i - P_{i,\max} \leq 0 \\ \frac{\partial \mathscr{L}}{\partial \mu_{i,\min}} = P_{i,\min} - P_i \leq 0 \\ \mu_{i,\max}(P_i - P_{i,\max}) = 0, \mu_{i,\max} \geq 0 \\ \mu_{i,\min}(P_{i,\min} - P_i) = 0, \mu_{i,\min} \geq 0 \end{cases} \tag{5.11}$$

Analyzing this set of equations, it is worth noting as $\mu_{i,\max} = 0$ when $P_i \leq P_{i,\max}$, and $\mu_{i,\min} = 0$ when $P_i \geq P_{i,\min}$. While, if the active power generated P_i falls outside the allowable operation limits (either $P_i \geq P_{i,\max}$ or $P_i \leq P_{i,\min}$), then it is constrained at that limit (either $P_i = P_{i,\max}$ or $P_i = P_{i,\min}$), and the ith generator is no longer available for the ED.

The solution of the problem formalized in (5.11) can be computed by deploying the following iterative scheme:

1. Define an initial candidate solution ($P_i = P_i^{(0)} \forall i \in [1, n_g]$, $\lambda = \lambda^{(0)}$.
2. Compute the gradient vector $\nabla \mathscr{L}|_x$ and the Hessian matrix Hx evaluated at the current solution point $x = [P_1, ...P_{n_g}, \lambda]$:

$$\nabla \mathscr{L}|_x = \begin{bmatrix} \frac{\partial C_1}{\partial P_1}|_x - \lambda + \frac{\partial P_L}{\partial P_1}|_x \\ \cdots \\ \frac{\partial C_{ng}}{\partial P_{ng}}|_x - \lambda + \frac{\partial P_L}{\partial P_{ng}}|_x \\ P_D + P_L - \sum_{i=1}^{n_g} P_i \end{bmatrix} \tag{5.12}$$

$$H|_x = \begin{bmatrix} \left.\frac{\partial^2 C_1}{\partial P_1^2}\right|_x + \left.\frac{\partial^2 P_L}{\partial P_1^2}\right|_x & 0 & \cdots & -1 \\ \cdots & \cdots & \cdots & \\ 0\ldots & 0 & \left.\frac{\partial^2 C_{ng}}{\partial P_{ng}^2}\right|_x + \left.\frac{\partial^2 P_L}{\partial P_{ng}^2}\right|_x & -1 \\ \frac{\partial P_L}{\partial P_1}|_x & \cdots & \frac{\partial P_L}{\partial P_{ng}}|_x & 0 \end{bmatrix} \tag{5.13}$$

3. Compute the variable variations by solving the following set of linear equations:

$$H_x[\Delta P_1, ..., \Delta P_{n_g} \Delta \lambda] = \nabla \mathscr{L}|_x \tag{5.14}$$

4. Adjourn the solution ($P_i^{(k+1)} = P_i^{(k)} - \Delta P_i \, \forall i \in [1, n_g]$, $\lambda^{(k)} = \lambda^{(k-1)} - \Delta \lambda$).
5. Verify the inequality constraints and properly update the list of dispatchable generators.
6. If a termination criteria is not met go to Step 1, otherwise end.

This solution scheme is frequently deployed in many power system operation tools, due to its reduced complexity, which is mainly influenced by the solution of the $n_g + 1$ linear equations (5.11).

To provide a clear explanation of this algorithm, let us examine a straightforward example involving the solution of the simplified ED problem formulated in (5.5). We will consider a power system with a power demand P_D of 1000 MW and three generators characterized by the following cost functions and operational limits:

$$\begin{cases} C_1(P_1) = 500 + 5.3P_1 + 0.004P_1^2 \; 0 \leq P_1 \leq 450\text{MW} \\ C_2(P_2) = 400 + 5.5P_2 + 0.006P_2^2 \; 0 \leq P_1 \leq 350\text{MW} \\ C_3(P_3) = 200 + 5.8P_3 + 0.009P_3^2 \; 0 \leq P_1 \leq 225\text{MW} \end{cases} \tag{5.15}$$

To solve the ED problem, we first compute the gradient vector and the Hessian matrix:

$$\nabla \mathscr{L}|_{P_1,P_2,P_3,\lambda} = \begin{bmatrix} 5.3 + 0.008P_1 - \lambda \\ 5.5 + 0.0012P_2 - \lambda \\ 5.8 + 0.0018P_3 - \lambda \\ 1000 - \sum_{i=1}^{3} P_i \end{bmatrix} \tag{5.16}$$

$$H|_x = \begin{bmatrix} 0.008 & 0 & 0 & -1 \\ 0 & 0.012 & 0 & -1 \\ 0 & 0 & 0.018 & -1 \\ -1 & -1 & -1 & 0 \end{bmatrix} \tag{5.17}$$

Then we start with a first estimation of the problem solution:

$$[P_1^{(0)}, P_2^{(0)}, P_3^{(0)}, \lambda^{(0)}] = [0, 0, 0, 10] \tag{5.18}$$

Hence, after implementing the first iteration of the solution algorithm, we get

$$\nabla \mathscr{L}|_{P_1^{(0)},P_2^{(0)},P_3^{(0)},\lambda^{(0)}} = \begin{bmatrix} -4.7 \\ -4.5 \\ -4.2 \\ 1000 \end{bmatrix} \tag{5.19}$$

$$[\Delta P_1^{(0)}, \Delta P_2^{(0)}, \Delta P_3^{(0)}, \Delta \lambda^{(0)}] = H^{-1} \nabla \mathscr{L}|_{P_1^{(0)},P_2^{(0)},P_3^{(0)},\lambda^{(0)}} = \begin{bmatrix} -494.74 \\ -313.16 \\ -192.10 \\ 0.74 \end{bmatrix} \tag{5.20}$$

$$[P_1^{(1)}, P_2^{(1)}, P_3^{(1)}, \lambda^{(1)}] = \begin{bmatrix} 494.74 \\ 313.16 \\ 192.10 \\ 9.26 \end{bmatrix} \tag{5.21}$$

Although the computed solution $[P_1^{(1)}, P_2^{(1)}, P_3^{(1)}, \lambda^{(1)}]$ satisfies the power system demand, namely

$$P_1^{(1)}, P_2^{(1)}, P_3^{(1)} = 1000\,\text{MW} \tag{5.22}$$

the output of the first generator is outside its allowable upper limit, namely $P_1 = 494.74 > 450\,\text{MW}$. Hence, according to the described solution scheme, we force the active power generated by the first generator to its maximum limit, namely

$P_1 = 450\,\text{MW}$, and we iterate the solution algorithm by considering only P_2 and P_3 available for the power dispatch: $P_2^{(1)} = 0$ $P_3^{(1)} = 0$. The second iteration of the solution algorithm allows computing the following quantities:

$$\nabla\mathscr{L}\big|_{P_2^{(1)},P_3^{(1)},\lambda} = \begin{bmatrix} -4.5 \\ -4.2 \\ 550 \end{bmatrix} \tag{5.23}$$

$$H|_x = \begin{bmatrix} 0.012 & 0 & -1 \\ 0 & 0.018 & -1 \\ -1 & -1 & 0 \end{bmatrix} \tag{5.24}$$

$$\left[\Delta P_2^{(2)}, \Delta P_3^{(2)}, \Delta\lambda^{(2)}\right] = H^{-1}\nabla\mathscr{L}\big|_{P_2^{(1)},P_3^{(1)},\lambda^{(1)}} = \begin{bmatrix} -340 \\ -210 \\ 0.42 \end{bmatrix} \tag{5.25}$$

$$\left[P_2^{(2)}, P_3^{(2)}, \lambda^{(2)}\right] = \begin{bmatrix} 340 \\ 210 \\ 9.58 \end{bmatrix} \tag{5.26}$$

which satisfies both the equality and inequality constraints, hence, representing the problem solution.

5.4 Solving the ED problem by a self-organizing sensors network

The deployment of the consensus protocols described in Chapter 2 allows a self-organizing agents network to effectively solve the ED problem formalized in (5.4) and (5.5) by a distributed/non-hierarchical solution scheme.

For this purpose, the first problem to solve is the estimation of the power system demand, which can be obtained by allowing each agent to assess the local power demand P_{D_i}, and to initialize the state of its oscillator as follows:

$$\omega_i = nP_{D_i} \quad \forall i \in [1,n] \tag{5.27}$$

This choice allows the agents network to self-synchronizing on the total grid power demand P_D:

$$\dot{\theta}^* = \sum_{i=1}^{n} P_{D_i} \tag{5.28}$$

which can be used by each agent to update the active power output of its controlled generating unit by using decentralized consensus protocols. In order to achieve this, let us first delve into the simplified formalization of ED analysis, as defined in (5.4). In particular, by considering (5.7), it results

$$\begin{cases} \Delta P_i = \dfrac{\frac{\partial C_i}{\partial P_i}|_x + \Delta\lambda - \lambda}{\frac{\partial^2 C_i}{\partial P_i^2}|_x} \quad \forall i \in [1, n_g] \\ -\sum_{i=1}^{n_g} \Delta P_i = P_D - \sum_{i=1}^{n} P_i \end{cases} \tag{5.29}$$

These equations can be combined as follows:

$$\sum_{i=1}^{n_g} \frac{\frac{\partial C_i}{\partial P_i}|_x + \Delta\lambda - \lambda}{\frac{\partial^2 C_i}{\partial P_i^2}|_x} = \sum_{i=1}^{n} P_i - P_D \tag{5.30}$$

These equations can be solved in order to determine $\lambda - \Delta\lambda$:

$$\lambda - \Delta\lambda = \frac{\sum_{i=1}^{n_g} \frac{\frac{\partial C_i}{\partial P_i}|_x}{\frac{\partial^2 C_i}{\partial P_i^2}|_x} + \Delta P}{\sum_{i=1}^{n_g} \frac{1}{\frac{\partial^2 C_i}{\partial P_i^2}|_x}} \tag{5.31}$$

where ΔP represents the power mismatch, which is defined as

$$\Delta P = P_D - \sum_{i=1}^{n} P_i = \sum_{i=1}^{n} P_{D,i} - P_i \tag{5.32}$$

This allows rearranging (5.31) as follows:

$$\lambda - \Delta\lambda = \frac{\sum_{i=1}^{n_g} \frac{1}{\frac{\partial^2 C_i}{\partial P_i^2}|_x} \left(\frac{\partial C_i}{\partial P_i}|_x + \frac{\Delta P}{n_g} \frac{\partial^2 C_i}{\partial P_i^2}|_x \right)}{\sum_{i=1}^{n_g} \frac{1}{\frac{\partial^2 C_i}{\partial P_i^2}|_x}} \tag{5.33}$$

hence, resulting in

$$\lambda - \Delta\lambda = \frac{\sum_{i=1}^{n_g} c_i \omega_i}{\sum_{i=1}^{n_g} c_i} \tag{5.34}$$

where

$$\begin{cases} \omega_i = \frac{\partial C_i}{\partial P_i}|_x + \frac{\Delta P}{n_g} \frac{\partial^2 C_i}{\partial P_i^2)}|_x \\ c_i = \frac{1}{\frac{\partial^2 C_i}{\partial P_i^2}|_x} \end{cases} \tag{5.35}$$

This important result allows each agent to infer the variable $\lambda - \Delta\lambda$ by deploying an average consensus protocol using the parameters settings defined in (5.35).

Once the agents network reaches a consensus on $\lambda - \Delta\lambda$, they can update the generator outputs as follows:

$$\Delta P_i = \frac{\frac{\partial C_i}{\partial P_i}|_x + \Delta\lambda - \lambda}{\frac{\partial^2 C_i}{\partial P_i^2}|_x} \quad \Delta P_i = P_i + \Delta P \, \forall i \in [1, n_g] \tag{5.36}$$

The same decentralized solution scheme can be properly enhanced in the task of considering the effects of the generator limits, and the active power system losses approximated by using (5.8) and (5.9). In particular, by according to the described solution scheme it follows that

$$\begin{cases} \Delta P_i = \dfrac{\frac{\partial C_i}{\partial P_i}|_x + 2B_i P_i + \Delta\lambda - \lambda}{\frac{\partial^2 C_i}{\partial P_i^2}|_x + 2B_i} \\ \sum_{i=1}^{n_g} (2B_i P_i - 1)\Delta P_i = P_D + P_L - \sum_{i=1}^{n} P_i \end{cases} \tag{5.37}$$

These equations allow inferring $\lambda - \Delta\lambda$ as follows:

$$\Delta\lambda - \lambda = \frac{\sum_{i=1}^{n_g} \dfrac{(2B_i P_i - 1)}{\frac{\partial^2 C_i}{\partial P_i^2}|_x + 2B_i} \left(\dfrac{\frac{\partial^2 C_i}{\partial P_i^2}|_x + 2B_i}{2B_i P_i - 1} \dfrac{\Delta P}{n_g} - \dfrac{\partial C_i}{\partial P_i}|_x + 2B_i P_i \right)}{\sum_{i=1}^{n_g} \dfrac{(2B_i P_i - 1)}{\frac{\partial^2 C_i}{\partial P_i^2}|_x + 2B_i}} \tag{5.38}$$

where the power mismatch ΔP can now be expressed as

$$\Delta P = P_D + P_L - \sum_{i=1}^{n} P_i = \sum_{i=1}^{n} (P_{D,i} - P_i + B_i P_i^2) \tag{5.39}$$

This allows recasting (5.38) as

$$\lambda - \Delta\lambda = \frac{\sum_{i=1}^{n_g} c_i \omega_i}{\sum_{i=1}^{n_g} c_i} \tag{5.40}$$

where

$$\begin{cases} \omega_i = \dfrac{\frac{\partial^2 C_i}{\partial P_i^2}|_{x+2B_i}}{2B_i P_i - 1} \dfrac{\Delta P}{n_g} - \dfrac{\partial C_i}{\partial P_i}|_x + 2B_i P_i \\ c_i = \dfrac{2B_i P_i}{\frac{\partial^2 C_i}{\partial P_i^2}|_x + 2B_i} \end{cases} \tag{5.41}$$

Hence, the agents network can infer the variable $\lambda - \Delta\lambda$ by deploying an average consensus protocol, using the parameters set described in (5.41). After self-synchronizing on this value, each agent can update its solution according to the following equation:

$$\Delta P_i = \frac{\frac{\partial C_i}{\partial P_i}|_x + 2B_i P_i + \Delta\lambda - \lambda}{\frac{\partial^2 C_i}{\partial P_i^2}|_x + 2B_i} \tag{5.42}$$

These results allow solving the ED problem by deploying the following decentralized solution scheme:

1. Initialize the variables.
2. Solve the average consensus problem by using the observation vector defined in (5.27) in the task of computing the power system load demand P_D.
3. Solve the average consensus problem formalized in (5.41) in the task of inferring the variable $\lambda - \Delta\lambda$.
4. Compute ΔP_i by using (5.42) and update the solutions $P_i = P_i + \Delta P_i \; \forall i \in [1, n_g]$.
5. Verify the inequality constraints and update the set of dispatchable generators.
6. Solve the following average consensus problem in the task of computing the power mismatch ΔP:

$$\omega_i = n\left(P_{D,i} - P_i + B_i P_i^2\right) \;\; \forall i \in [1, n_g] \tag{5.43}$$

7. Check for convergence: if the termination criteria is not met go to Step 2, otherwise, end.

In order to clearly illustrate this solution scheme, let us solve again the 3-generators example described in Section 5.3. For this purpose, let us start from the same initial solution:

$$x^{(0)} = \left[P_1^{(0)}, P_2^{(0)}, P_3^{(0)}\right] = [0,0,0] \tag{5.44}$$

In the initial phase, the dynamic agents calculate the power system demand through the solution of an average consensus problem, which allows them to infer the variable P_D by exclusively exchanging local data, effectively bypassing the need for a central fusion center. Upon the network of agents achieving synchronization, the state of each agent converges toward the actual value of P_D, which is 1000 MW.

Following this, the dynamic agents undertake the task of addressing an average consensus problem to compute the global variable $\lambda^{(0)} - \Delta\lambda^{(0)}$. In this scenario, as the agents network attains synchronization, the state of each agent converges to the subsequent values:

$$\lambda^{(0)} - \Delta\lambda^{(0)} = \frac{\Sigma_{i=1}^{3} \frac{1}{\frac{\partial^2 C_i}{\partial P_i^2}|_x^{(0)}} \left(\frac{\partial C_i}{\partial P_i}|_x^{(0)} + \frac{\Delta P}{3} \frac{\partial^2 C_i}{\partial P_i^2}|_x^{(0)} \right)}{\Sigma_{i=1}^{3} \frac{1}{\frac{\partial^2 C_i}{\partial P_i^2}|_x^{(0)}}}$$
$$= \frac{\frac{5.3 + 0.008 P_1^{(0)}}{0.008} + \frac{5.5 + 0.012 P_2^{(0)}}{0.012} + \frac{5.8 + 0.018 P_3^{(0)}}{0.018} + 1000}{\frac{1}{0.008} + \frac{1}{0.012} + \frac{1}{0.018}} = 9.26 \tag{5.45}$$

This is confirmed in Figure 5.1, which reports the time evolution of the derivative of the agents state synchronizing on $\lambda^{(0)} - \Delta\lambda^{(0)}$.

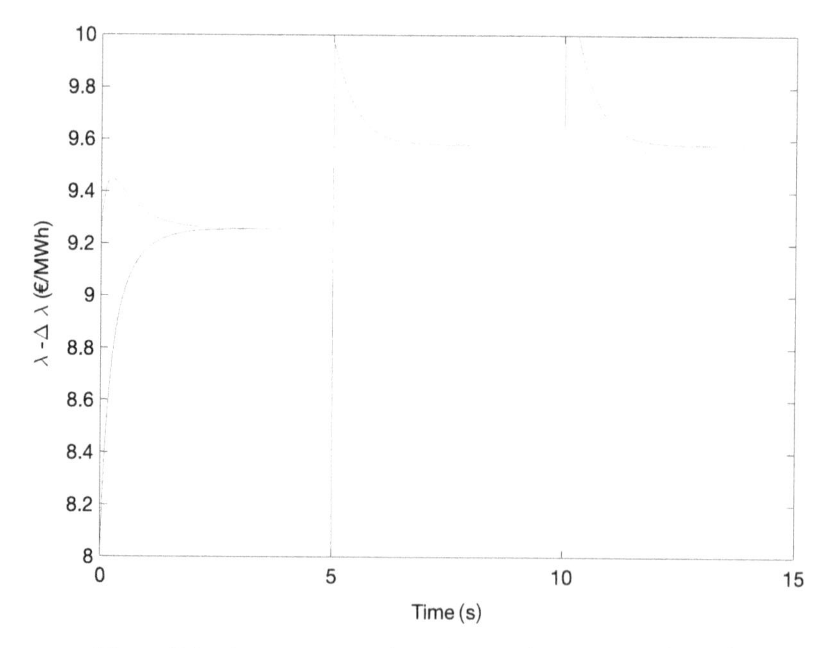

Figure 5.1 Agents state trajectories synchronizing on $\lambda - \Delta\lambda$

After synchronizing on $\lambda^{(0)} - \Delta\lambda^{(0)}$, the agents network determines the corresponding $\Delta P_i^{(0)}$, and updates the problem solution, as follows:

$$\begin{aligned}
\Delta P_1^{(0)} &= \frac{\frac{\partial C_1}{\partial P_1}|_x^{(0)} + \Delta\lambda^{(0)} - \lambda^{(0)}}{\frac{\partial^2 C_1}{\partial P_1^2}|_x^{(0)}} \\
&= \frac{5.3 + 0.008 P_1^{(0)} - 9.26}{0.008} = -495 \\
\Delta P_2^{(0)} &= \frac{\frac{\partial C_2}{\partial P_2}|_x^{(0)} + \Delta\lambda^{(0)} - \lambda^{(0)}}{\frac{\partial^2 C_2}{\partial P_2^2}|_x^{(0)}} \\
&= \frac{5.5 + 0.012 P_2^{(0)} - 9.26}{0.012} = -313.3 \\
\Delta P_3^{(0)} &= \frac{\frac{\partial C_3}{\partial P_3}|_x^{(0)} + \Delta\lambda^{(0)} - \lambda^{(0)}}{\frac{\partial^2 C_3}{\partial P_3^2}|_x^{(0)}} \\
&= \frac{5.8 + 0.018 P_3^{(0)} - 9.26}{0.018} = -192.2
\end{aligned} \tag{5.46}$$

$$\begin{cases} P_1^{(1)} = P_1^{(0)} - \Delta P_1^{(0)} = 495 \\ P_2^{(1)} = P_2^{(0)} - \Delta P_2^{(0)} = 313.3 \\ P_3^{(1)} = P_3^{(0)} - \Delta P_3^{(0)} = 192.2 \end{cases} \tag{5.47}$$

At this stage, Agent 1 identifies that the power generated by its respective unit is greater than its upper threshold ($P_1 > 450$ MW). Consequently, it constrains its output to this limit of 450 MW, thereby opting out of the dispatch process.

Subsequently, the agents network collaborates to compute the corresponding power imbalance denoted as $\Delta P(0)$, achieved by solving an average consensus problem. The outcomes of this computation are depicted in Figure 5.2. Upon closer examination of the illustration, it is worth observing as during the second main iteration, the agents network self-synchronizes on the value of −44.5 MW, which is the actual power imbalance ($\Sigma_{i=1}^{3} P_i^{(0)} - 1000$).

After synchronizing on the actual power imbalance, the agents controlling the generator units still available for dispatching reach an average consensus on $\lambda^{(1)} - \Delta\lambda^{(1)}$. In particular, the analysis of Figure 5.1 confirms that network synchronizes on the correct value:

$$\lambda^{(1)} - \Delta\lambda^{(1)} = 9.58 \tag{5.48}$$

Once reached this synchronization state, each agent determines the corresponding $\Delta P_i^{(1)}$ and updates its solution as follows:

$$\begin{cases} \Delta P_2^{(1)} = -26.70 \\ \Delta P_3^{(1)} = -17.80 \end{cases} \tag{5.49}$$

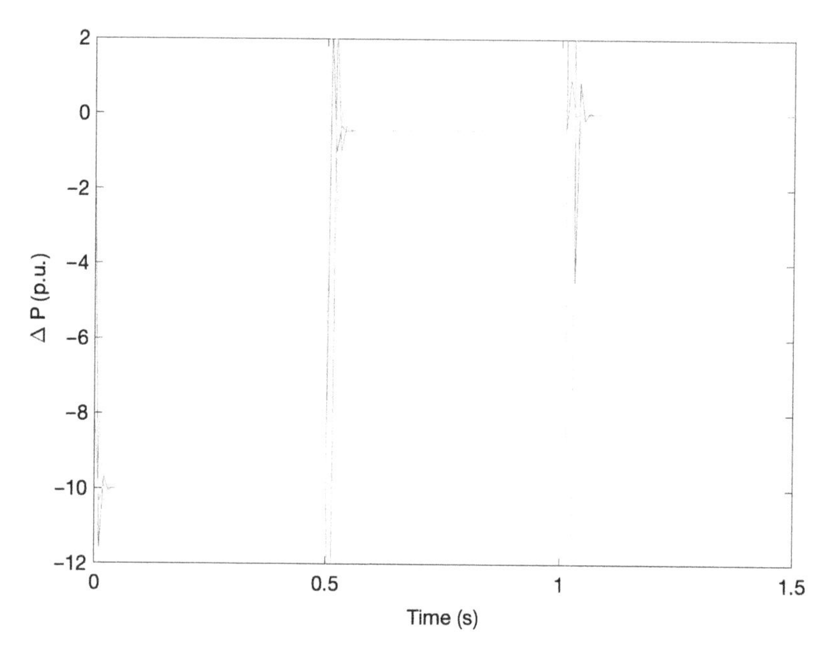

Figure 5.2 Agents state trajectories synchronizing on ΔP assuming a base power of 100 MVA

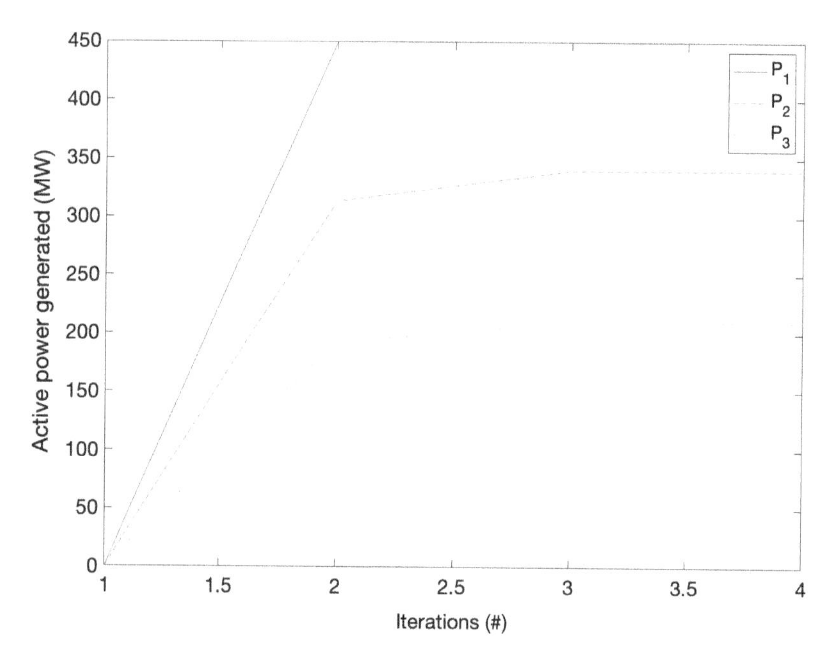

Figure 5.3 ED solution computed by the self-organizing agents network

$$\begin{cases} P_1^{(2)} = 450 \\ P_2^{(2)} = 340 \\ P_3^{(2)} = 210 \end{cases} \tag{5.50}$$

Finally, the agents network updates the power imbalance estimation $\Delta P^{(1)}$, and, considering the null active power balance, and the feasibility of the computed solution to all the inequality constraints, the algorithm terminates.

This illustrative example underscores the noteworthy observation that the decentralized algorithm converges toward the same solution as obtained through the application of the conventional algorithm outlined in Section 5.3. This observation stems from the fact that both algorithms originate from the same mathematical core, which is based on computing the stationary points of the Lagrangian function.

In terms of algorithm complexity of the decentralized framework, it is primarily dependent on the number of iterations required for self-synchronization, which is significantly influenced by the coupling functions and the logical graph defining the interconnections among the dynamic agents.

5.5 Example

Within this section, an evaluation of the effectiveness of the decentralized solution framework is conducted in addressing the ED problem for a network of 30 dispatchable generators, each defined by quadratic cost function. For the purpose of this analysis, the system power demand has been fixed to 1000 MW, and the cost coefficients are assumed to vary within the subsequent ranges:

$$\begin{cases} \alpha_i \in [4.5, 6] \\ \beta_i \in [0.004, 0.009] \\ \alpha_i \in [200, 500] \ \forall i \in [1, 30] \end{cases} \tag{5.51}$$

while the minimum and the maximum allowable power generated by each generator unit have been fixed as

$$10 \leq P_i \leq 73\, MW \ \forall i \in [1, 30] \tag{5.52}$$

For the sake of simplicity, the impact of system losses has been disregarded. This decision does not compromise the validity of the described methodology since the network loss modeling does not alter the fundamental mathematical foundation of the decentralized solution scheme.

A dynamic agents network comprising of 30 units, each associated with a power generation unit, has been deployed across the power system.

The preliminary phase of our investigation revolves around evaluating the dynamic agents ability to accurately assess the actual power system demand. The outcomes of this stage are depicted in Figure 5.4. As expected, the application of the distributed average consensus protocol enables the synchronization of all agents (initial states chosen randomly) toward the true value of the power demand. This

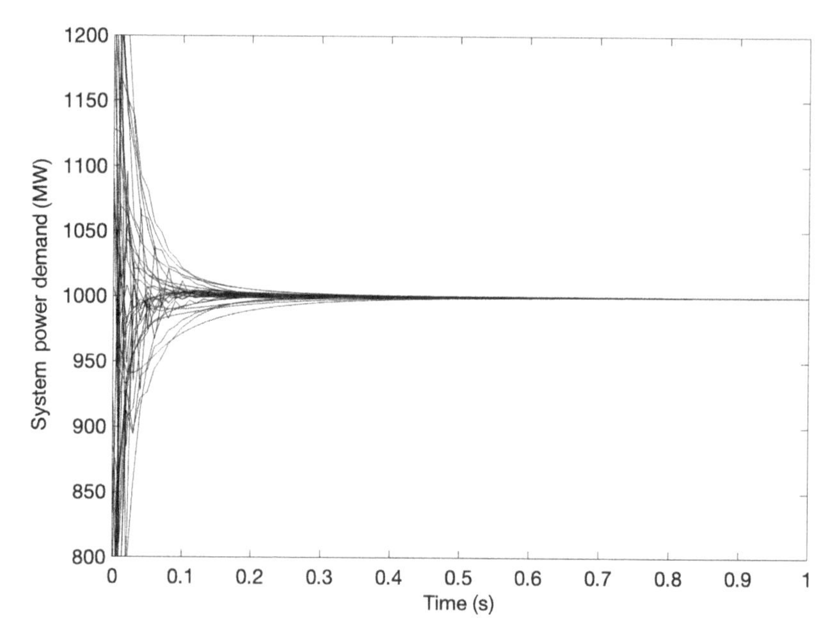

Figure 5.4 Agents state trajectories synchronizing on the system power demand

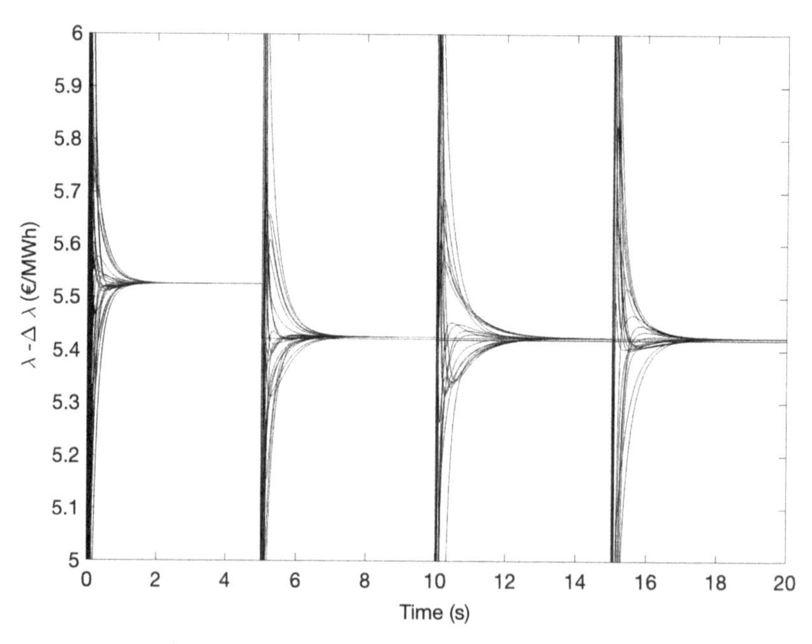

Figure 5.5 Agents state trajectories synchronizing on $\lambda - \Delta\lambda$

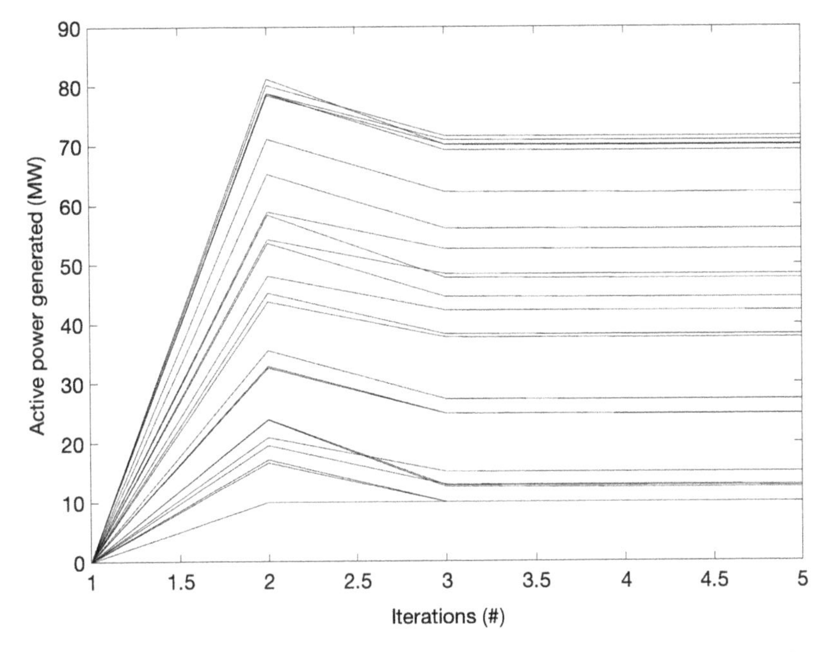

Figure 5.6 ED solution computed by the self-organizing agents network

convergence is achieved in approximately 1 s. It is important to note that the incorporation of an enhanced non-linear coupling protocol is expected to substantially reduce the required convergence iterations.

The next phase of our exploration focuses on evaluating the performance of the agents network in solving the ED problem. The results obtained are summarized in Figures 5.4 and 5.5, which report the trajectories of the agents states synchronizing on $\lambda - \Delta\lambda$ and ΔP, respectively. Upon analyzing these trajectories, it is apparent that the dynamic agents reach a consensus on the real values of the variables of interest. Furthermore, they infer the optimal dispatch solution within just three control iterations.

This conclusion is further substantiated upon reviewing Figure 5.6, which outlines the progression of the solutions P_i identified by the individual agents. These profiles allow for an evaluation of the influence of maximum generation limits on the computed solution. Notably, after the second control iteration, it becomes evident that many generators are constrained to operate at their maximum allowable limits, excluding them from further participation in the optimization process.

Turning to the practical viability of the proposed computational paradigm, several key considerations emerge:

1. The computational burden primarily stems from solving a set of mutually coupled first-order differential equations, whose convergence hinges on the eigenvalues of the adjacency matrix describing the logical connections between the dynamic agents. This insight enables a theoretical estimation of the impact of communication delays and fluctuating availability of communication links on

algorithm stability and convergence according to the methodologies described in Chapter 2.

2. As rigorously established in [16], the decentralized consensus protocols attain convergence in communication networks characterized by dynamically changing topologies, as long as the communication graphs maintain joint connectivity.
3. The consensus paradigm deployed in this example exhibits a propensity for convergence towards average consensus, even in the presence of random communication link failures.

These attributes hold considerable promise for the seamless and efficacious implementation of the described decentralized architecture within modern power systems tools.

References

[1] Vlachogiannis JG and Lee KY. Economic load dispatch—a comparative study on heuristic optimization techniques with an improved coordinated aggregation-based PSO. *IEEE Transactions on Power Systems*. 2009;24(2):991–1001.

[2] Reddy YVK, Devi ML, Reddy AVS, *et al*. Economic dispatch solutions with piecewise quadratic cost functions using spotted hyena optimizer. In: *2021 IEEE Madras Section Conference (MASCON)*; 2021. p. 1–6.

[3] Tashiro T, Tamura K, and Yasuda K. Modeling and optimal operation of distributed energy systems via dynamic programming. In: *2011 IEEE International Conference on Systems, Man, and Cybernetics*; 2011. p. 808–813.

[4] Sinha N, Chakrabarti R, and Chattopadhyay PK. Evolutionary programming techniques for economic load dispatch. *IEEE Transactions on Evolutionary Computation*. 2003;7(1):83–94.

[5] Sharma U and Moses B. Analysis and optimization of economic load dispatch using soft computing techniques. In: *2016 International Conference on Electrical, Electronics, and Optimization Techniques (ICEEOT)*; 2016. p. 4035–4040.

[6] Mudumbai R, Dasgupta S, and Cho B. Distributed control for optimal economic dispatch of power generators. In: *Proceedings of the 29th Chinese Control Conference*; 2010. p. 4943–4947.

[7] Essakiappan S, Shoubaki E, Koerner M, *et al*. Dispatchable virtual power plants with forecasting and decentralized control, for high levels of distributed energy resources grid penetration. In: *2017 IEEE 8th International Symposium on Power Electronics for Distributed Generation Systems (PEDG)*; 2017. p. 1–8.

[8] Balachennaiah P and Chinnababu J. Distributed control based multi agent system for energy management of hybrid microgrids system. In: *2022 International Conference on Emerging Trends in Engineering and Medical Sciences (ICETEMS)*; 2022. p. 296–301.

[9] Loia V and Vaccaro A. Decentralized economic dispatch in smart grids by self-organizing dynamic agents. *IEEE Transactions on Systems, Man, and Cybernetics: Systems*. 2014;44(4):397–408.

[10] Mohsenian-Rad AH and Leon-Garcia A. Optimal residential load control with price prediction in real-time electricity pricing environments. *IEEE Transactions on Smart Grid*. 2010;1(2):120–133.

[11] Tang R, Wang S, and Li H. Game theory based interactive demand side management responding to dynamic pricing in price-based demand response of smart grids. *Applied Energy*. 2019;250:118–130.

[12] Kishore S and Snyder LV. Control mechanisms for residential electricity demand in smart grids. In: *2010 First IEEE International Conference on Smart Grid Communications*. IEEE; 2010. p. 443–448.

[13] Hajibandeh N, Shafie-Khah M, Osório GJ, *et al.* A heuristic multi-objective multi-criteria demand response planning in a system with high penetration of wind power generators. *Applied Energy*. 2018;212:721–732.

[14] Gao Dc, Sun Y, and Lu Y. A robust demand response control of commercial buildings for smart grid under load prediction uncertainty. *Energy*. 2015;93:275–283.

[15] Wang Y, Liang H, and Dinavahi V. Two-stage stochastic demand response in smart grid considering random appliance usage patterns. *IET Generation, Transmission & Distribution*. 2018;12(18):4163–4171.

[16] Patterson S, Bamieh B, and El Abbadi A. Convergence rates of distributed average consensus with stochastic link failures. *IEEE Transactions on Automatic Control*. 2010;55(4):880–892.

Chapter 6

Decentralized monitoring estimation and control of smart microgrids

From a conceptual perspective, a microgrid (MG) is a distribution system operating at either medium- or low-voltage levels, integrating an array of dispersed generators and storage units [1,2]. The integration of MGs has the potential to enhance the technical, economic, and environmental aspects of conventional power distribution systems through the following avenues:

1. Facilitating widespread integration of distributed generators.
2. Supporting the integration of renewable power sources.
3. Curtailing power losses and emissions.
4. Enhancing power quality and system reliability [3].

Nonetheless, the realization of these benefits necessitates the solution of various challenges. For instance, as expounded in [4], the control and management paradigms of MGs can significantly deviate from the conventional architectures employed in conventional power distribution systems. These differences primarily arise from factors such as the nature and penetration levels of distributed power generators, the presence of controllable loads, the definition of strict power quality criteria, and the randomness derived from the complex dynamics ruling the electricity markets.

Modern frameworks used for MG control and management are underpinned by networks of distributed and heterogeneous sensors/controllers, which gather and process data from dispersed energy resources and can be deployed using centralized, decentralized, or hybrid computing architectures [2]. In centralized architectures, field data are collected by a distributed metering system and transmitted to central servers, which process these data in the task of determining proper control actions aimed at optimizing MG operation on the basis of the actual demand, environmental conditions, energy market prices, and network constraints. The resulting control settings are then transmitted back to the primary controllers of the dispatchable generators and flexible loads [5]. Conversely, the decentralized control paradigm seeks to bestow dispersed generators and load entities with maximal decision-making autonomy. As a consequence, local controllers must engage in communication and coordination with one another. This collective control network culminates in an intelligent entity often referred to as a smart microgrid (SMG) [3]. In this context, the primary objective of each entity is to enhance the overall performance of the SMG, rather than merely maximizing the performance of individual units. Thus, the control architecture should possess the capability to estimate key variables characterizing SMG operation by solely processing local data. These characteristics suggest that the multi-agent system (MAS) paradigm offers a promising avenue for decentralized SMG control and management [6]. The advantages stemming from the application of MAS in SMG control and management encompass [7]:

1. Effective decentralized data processing and decision-making.
2. High scalability and adaptability.
3. Self-healing and self-organization capabilities.
4. Knowledge extraction from large data-sets.

In this context, the work presented in [6] proposes a comprehensive framework for managing distributed energy resources within the SMG domain. This architecture, founded on the MAS approach, endeavors to integrate various functionalities while accommodating the SMG complexity. Achieving this adaptability involves employing a layered learning paradigm that organizes diverse control loops and agent actions based on their impact on global grid performances.

The effectiveness of deploying agent-based paradigms in SMG computing is underscored in [8]. This work showcases how cooperative smart agents, aligned with interoperable protocols, represent a pragmatic and viable methodology for enhancing operational efficiency within SMGs. A decentralized MAS architecture designed to address voltage and frequency control within SMGs is introduced in [9]. Differently from conventional primary voltage/frequency control, this architecture hinges on information-spreading paradigms orchestrated by a network of cooperative agents. Further studies reiterate the advantages derived from applying MAS in solving varied SMG challenges, spanning load management, economic dispatch, emergency control, pervasive monitoring, and network modeling [10].

While these studies contribute valuable insights into the significance of MAS in the context of SMGs, they predominantly address specific control functionalities. Moreover, they lack a holistic, integrated, and decentralized framework that can tackle multiple control and monitoring tasks, while adhering to standardized and effective implementation. Furthermore, they do not explicitly tackle a critical issue in synchronized SMG control applications—the vulnerability of the time synchronization source. In particular, although satellite-based timing signals provide highly accurate global time references for synchronizing sensors and controllers, they remain susceptible to radiofrequency interference (RFI). Consequently, effective countermeasures aimed at enhancing the resilience of synchronized SMGs to external and internal RFI stand as crucial challenges.

In this chapter, we describe the utilization of self-organizing sensor networks equipped with the decentralized consensus protocols described in Chapter 2 in the task of addressing some of the fundamental challenges characterizing modern SMG operation. Specifically, we describe the decentralization of monitoring, state estimation, control, and synchronization functions within an SMG. To accomplish this, we deploy a network of dynamic and self-organizing agents equipped with distributed consensus protocols. This approach demonstrates how consensus protocols enable dynamic agent networks to acquire global information, facilitating interaction and cooperation in the execution of complex tasks. This is achieved without the necessity of fixed message-passing protocols or routing schemes. This characteristic empowers dynamic agents to solve fundamental SMG control and monitoring issues in a fully decentralized manner, without the requirement for central facilities to process all SMG field data.

Furthermore, agents can achieve time synchronization by compensating for sensor clock drift through local coupling strategies, obviating the need for any cluster header. This synchronization mechanism allows dynamic agent clocks to synchronize their phases despite frequency variations. This redundancy ensures dynamic agents continue to function effectively in situations of main SMG synchronization source unavailability or degradation.

To validate the performance of this self-organizing framework, we present and discuss simulation results obtained by implementing the dynamic agent-based paradigm in the control, monitoring, time synchronization, and estimation of an 18-bus SMG.

6.1 Enabling technologies for enhancing smart grids flexibility

6.1.1 Virtual power plants

Renewable energy sources like wind and solar power generators cannot provide grid flexibility as individual units, as their power output cannot be precisely controlled. However, when multiple renewable energy plants are combined and properly operated in conjunction with dispatchable generators and energy storage systems, they can be modeled as a unified energy entity offering flexible services. This is the fundamental concept behind a virtual power plant (VPP) [11].

VPP is a modern concept that allows for the management of distributed energy resources in a coordinated and flexible way. It is essentially an aggregation of various generating units with different characteristics, such as dispatchable and renewable power generators, flexible loads, and energy storage units. This aggregation allows for the smart coordination of these components, such that specific objective functions (e.g. economic profits and environmental costs) are maximized/minimized. The VPP combines the power produced by various generators, stored by batteries, or absorbed by flexible loads in such a way that it can sell energy and services to the electricity market.

From a market perspective, a VPP is viewed as a single flexible power output that manages its generation and demand internally, thereby reducing variability and adding a source of flexibility to the power system.

The concept of VPP has gained significant attention in the power system domain in recent times. VPPs are characterized by three main features, namely, distributed information collection and processing, wide-area geographic distribution, and dynamic scheduling. By collecting real-time information on the available components, VPPs can optimize the scheduling of each energy resource [12]. Moreover, a VPP can be geographically dispersed, and its resources need not be connected to the same bus of the power distribution grid. VPPs also allow real-time scheduling of their components, enhancing decision-making processes.

Broadly, VPPs can be classified into two types, namely commercial and technical. A commercial VPP focuses on obtaining the largest economic profit from the electricity markets, whereas a technical VPP aims at ensuring the safe operation of VPP and improving the power system balance or offering grid ancillary services.

VPPs can be controlled using three different paradigms: centralized control, distributed control, and hybrid control. In centralized control, there is a control center that dispatches all of the available energy resources. This kind of control allows the VPP to present itself to the electricity market as a price-maker instead of a price-taker, given the large amount of power it can produce. The centralized problem can be solved using a MILP problem, considering stochastic or robust counterparts if uncertainty has to be taken into account [13]. However, the main drawback of centralization is its limited scalability, due to the computational requirements for the optimization of large systems.

The approach of distributed optimization involves the cooperation of groups of resources that aim to achieve a common objective, such as maximum profit, minimum emissions, or other specific targets [14]. Game theory is often employed to enable distributed control in this approach.

On the other hand, the hybrid control method combines the features of both distributed and central control. It comprises two levels – a centralized one, which determines the final bidding strategy for market participation, and a distributed level, which optimizes the local resources regionally. This approach is particularly useful in handling geographically dispersed VPPs.

According to [15], a VPP can be divided into four main components: dispatchable generators, flexible loads, energy storage systems, and not-programmable generation units.

Dispatchable generators are small-scale generating units fueled by fossil fuels or biomass sources, which are typically used to cover the base load of a VPP. An example of this kind of generator is the cogenerator or trigenerator, which allows generating electricity, heat, and cooling.

Flexible loads include residential or industrial consumers who are equipped with communication and monitoring systems, which allow them to adjust their power demand profiles based on control signals from the central VPP controller. This makes it possible to balance the load and generation in real-time, improve the grid stability, and reduce the need for peak power plants.

Energy storage systems are important components of VPP, as they allow managing the intermittence of renewable power generators, mitigating their grid impacts, while ensuring reliable and stable power system operation.

Not-programmable generation units include renewable power generators, whose operation strictly relies on weather conditions, such as wind turbines and solar generators. The power profiles generated by these units are inherently unpredictable, as they are subject to fluctuations in weather patterns. However, by aggregating a large number of renewable power generators in a VPP, it is possible to reduce the variability of the overall output, enhancing the system predictability.

The optimal management strategy for a VPP can be modeled through an optimization problem with constraints imposed by each component of the VPP, as explained in [16]. By optimizing the dispatch of all the VPP components, it is possible to reduce the cost of energy and improve the efficiency of the system.

6.1.1.1 Multi-energy systems

It is possible to enhancing the hosting capability of renewable energy sources in smart grids by coordinating multiple energy carriers. The power-to-gas paradigm is an example of such integration, which involves converting electric energy to hydrogen using electrolysis or to methane through methanation. The produced gas can be stored for future use, with an efficiency ranging from 60% to 80% [11].

However, storage losses can reduce the efficiency of this process, and this is still a topic of debate. Moreover, while hydrogen storage requires high-cost investments, methane can be absorbed by existing gas distribution systems or storages. Power-to-gas is expected to play a larger role in ancillary and spot electricity markets as it is integrated into regulations.

In terms of flexibility, integrating different energy systems has great potential. Multi-energy systems can optimize different energy vectors, including heat, electricity, and gas, thereby reducing costs, emissions, and providing a flexibility source.

The widespread adoption of combined heat and power generators, energy storage systems, and V2G-based electric vehicles, requires new computing frameworks aimed at modeling heterogeneous energy infrastructures and multiple energy carriers in an integrated manner. Rather than considering these networks as separate and independent, characterized by their own state variables, they should be viewed as a whole system [17]. In this context, the concept of energy loads should evolve from passive entities that only consume energy to dynamic and cooperative entities capable of providing flexible services to the available energy systems [18].

Multi-energy systems can be classified into four main categories based on different perspectives [19]. First, the spatial context considers that multi-energy systems can be aggregated at various geographical levels, ranging from individual loads to smart communities. Second, the multi-service perspective evaluates that multi-energy systems can supply different types of energy loads simultaneously (e.g. thermal and electrical). Third, the multi-fuel perspective highlights that the energy inputs may be of varying types, including fossil fuels, renewable energy sources, hydrogen, and heat. Finally, the network perspective emphasizes the cooperation between energy networks, such as power and natural gas, in order to optimize their operations in terms of cost and emissions.

In this perspective emerges the concept of an energy hub, which is analogous to an electric bus in networks.

It is widely acknowledged that the energy hub represents a promising framework to achieve a decarbonized, efficient, and distributed power system. An energy hub is a functional unit that can condition, store, and convert energy carriers. It is closely related to energy interconnectors, which are transmission elements that carry multiple energy flows through various nodes. These elements generalize the concepts of electric power networks buses and power lines to multi-carrier systems [20].

An energy hub is composed of a network of interconnected energy converters and storage systems that aim to process input energy carriers, such as electric energy, gas, and district heating, to supply electrical, thermal, and chemical loads. Typical components of energy hubs are combined heat and power systems, fuel cells, boilers, and others, which can be modeled at various scales, ranging from local loads [21] to large-scale systems [22].

Despite the fact that energy hubs can provide significant flexibility, there are still some open issues that need to be addressed in order to fully exploit their benefits. One of these issues is their economic evaluation in the presence of uncertainties, which can refer to both operation, that is optimal hub dispatch, and planning, i.e., optimal hub sizing.

To solve this problem, conventional methods formulate deterministic constrained optimization models [17,23]. However, input data are affected by a large number of uncertainty sources, which include the volatility of energy prices, random load fluctuations, the variability of energy production from renewable power generators, and the approximation errors affecting the energy hub models.

Although some attempts have been made to address uncertain dispatch problems through robust and stochastic optimization [24], these methods have limitations that restrict their practical use. In the case of sampling methods, the set of simulations can be extremely large and computationally expensive, while analytical techniques require a series of simplifying assumptions, such as an imprecise characterization of random variables in terms of their probability distribution [25]. Decision-makers may not always have access to the data needed for such characterization, and strong assumptions can lead to unrealistic outcomes.

6.1.2 Microgrids

A microgrid can be considered a network of electrical loads and distributed energy resources, which are appropriately coordinated to act as a single controllable flexible entity concerning the grid [26]. The primary resources that can be integrated into a microgrid can be categorized as follows:

1. Generation: The generation system can include a variety of dispatchable and non-dispatchable sources. Dispatchable options include natural gas generators, biogas generators, and combined heat and power, while non-dispatchable sources consist of renewable generators as far as solar, wind, and hydro are concerned.
2. Energy storage system: It may implement several essential functionalities, such as power quality enhancement, peak load shaving, primary/secondary frequency control, mitigating the randomness of renewable power generators, providing a redundant energy source for critical loads, and enhancing the microgrid flexibility.

3. Intelligent management system: It acquires and processes the microgrid data to schedule the available energy resources by dispatching the generators and the flexible loads according to economic or environmental criteria.
4. Electrical loads: They can be classified into two categories:
   (a) Critical loads, which should be supplied under all operating conditions.
   (b) Flexible loads, whose power demand can be adjusted to enable load balancing, enhance power generators economic operation.
5. Real-time controller: It manages the instantaneous microgrid operation, and its connection with the distribution grid.
6. Point of common coupling: It is an essential component for grid-connected microgrids, representing the physical connection between the microgrid and the power distribution system. It represents the external power grid interface, which integrates various control and protection components to enable the grid connection and active/reactive power exchange with the external grid. It comprises circuit breakers, protective systems, and synchronization equipment.

There are several potential benefits of microgrids that make them an attractive option for power supply. First, they offer price stability by hedging the risk of unforeseeable and potentially expensive contingency/emergency energy costs, and price spikes, hence protecting against fluctuating energy sourcing costs. Second, depending on electricity market schemes, microgrids can mitigate peak load costs, enable demand response programs, and provide frequency support services to distribution power systems, generating economic profits. Third, they ensure continuous power supply, especially in cases where the power distribution grid faces outages due to extreme weather, aging infrastructures, and physical or cyberattacks. In this scenario, autonomous microgrids (i.e., operating in island mode) can guarantee reliable electrical energy supply by separating from the external power grid and using the available generators and storage systems to supply the local loads. Fourth, microgrids represent an enabling technology to enhance the hosting capacity of renewable power generators by mitigating their adverse side effects on power grids.

Moreover, the capacity to operate in autonomous mode makes microgrids an effective solution for enhancing grid reliability and resilience against extreme perturbation phenomena, allowing them to supply power to critical electrical loads also in the presence of power outages and isolating faults by separating distribution feeds.

Additionally, implementing microgrid allows better control over power quality parameters, which is crucial for sensitive equipment in healthcare and critical manufacturing processes.

For these reasons, microgrids are considered the essential building element for smart grids, and future power distribution systems rely on dynamic networks of cooperative, interconnected microgrids aimed at managing energy demand and supply at the micro and macro levels.

In the context of microgrids, there are various use cases that are representative of the application domains. Institutional microgrids usually consist of multiple buildings within a confined geographical location. Depending on the institution nature, the reliability and power supply quality requirements may differ. For example, government facilities may be satisfied with moderate power supply reliability, whereas military institutes may require a higher quality power supply. In such microgrids, all participants belong to a single entity, and there is a single decision-maker, which allows for dynamic energy resource scheduling, and prompt action when economic benefits arise.

When a microgrid is established in an existing commercial or industrial area, the scenario becomes more complex. In this application domain, both premium and standard power supply capabilities can be defined for each participant, hence, system designers should identify a microgrid architecture that satisfies all the user requirements. The main goal is to diversify the energy source mix in order to minimize the energy sourcing costs, enhancing the energy self-sufficiency, and the supply security by optimal scheduling the local energy sources reacting to the spot-market price dynamics.

Private end-customers primarily residing in residential areas may develop "community and utility"-based microgrids, which can include urban regions, communities, and rural feeders and can be connected to the power distribution system, providing power to both urban and rural areas. The deployment of this type of microgrid, which can include a wide range of renewable or fossil-fueled distributed energy resources, can be boosted by national and international standards and regulations. However, the decision-making process and the microgrid design may take longer due to the large number of participants.

A community or utility microgrid is typically based on an island system, since this system is usually not connected to the main power grid although, in some cases, a grid connection may be possible if the distance from the main power distribution system allows it. The decision-making process could be expedited depending on the actual power supply infrastructure of the microgrid. For geographically isolated or remote area and developing countries, isolated microgrids, which are based on distributed and heterogenous power sources, are typically employed. While, remote microgrids could be implemented in the task of connecting to future power system, as in the case of developing world regions, which continue to improve their electrical infrastructures.

6.1.2.1 Microgrid classifications

To assist designers in adopting safe and robust design options for microgrids (MGs), a thorough literature analysis was carried out in [26] to explore the primary topologies and architectural structures of modern microgrids, which can be classified into various categories based on their applications, infrastructure, and end-user needs.

1. Microgrid control strategies:
   - Centralized control: In microgrids with centralized control, the central controller communicates with the local controllers via a two-way communication channel to provide optimal set points.
   - Decentralized control: On the other hand, MGs with decentralized control typically adopt a control technique based on multiagent systems. The operation of each microgrid component is individually defined and orchestrated, without the need for a centralized controller. These control architectures

offer flexibility in all the grid functions, and the component interactions can be obtained by using standard communication language such as Java-Jade.

2. Microgrid size:
   - Small-scale microgrids: they are capable of producing electrical energy through the use of renewable energy sources. However, some small-scale systems may also rely on conventional generators as an alternative or replacement source of power. These small-scale microgrids are typically capable of generating up to 10 MW and can provide electrical power to residential buildings, remote areas, and islands.
   - Medium-scale microgrids: they are designed to generate medium-capacity electrical energy and can use a combination of renewable energy sources, and conventional power generators. The generation capacity for medium-scale microgrids can range from more than 10 MW up to 100 MW. These MGs are best suited for supplying industrial loads.
   - Large-scale microgrids: they are capable of generating high-capacity electricity through the use of conventional generators. These systems typically have a generation capacity of over 100 MW and are best suited for supplying large industrial areas.
3. Classifying MG on the basis of power supply, in terms of connected power supply, MGs can be divided into three categories: AC, DC, and hybrid MGs [26].
   - AC microgrid: an AC microgrid is a system that comprises sinusoidal power generators, which supply AC loads. This system can operate isolated or connected to the primary power system at the point of common coupling. To meet the AC load requirements, the AC bus links all the local energy resources, namely the power generators, storage devices, and other system components. These microgrids are easy to integrate into existing power systems and do not require additional control mechanisms. They can be classified by the electrical supply scheme (single-phase, grounded three-phase, and ungrounded three-phase), or based on the supply frequency (high-frequency, low-frequency, and standard-frequency). In practical applications, AC microgrids are the most widely used architecture among other options. Nonetheless, synchronizing with the power distribution system while guaranteeing that voltage magnitude, phase angle, and frequency vary within allowable ranges is a problematic control issue to address.
   - DC microgrids, which store and generate electrical power in DC forms, have a supply power that follows DC power and drives connected loads with DC power. These microgrids offer several advantages over AC microgrids, including not requiring grid synchronization, rarely experiencing power quality problems, and not having all the problems related to power factor correction. However, DC microgrids use multiple converters and power electronic devices to interface with current distribution systems. DC microgrids have higher energy conversion efficiency when directly supplying DC loads and are used in various commercial applications, such as telecommunication, electric vehicles, and marine power systems. DC microgrids allow DC loads to connect directly to the DC bus, which reduces the need for many

power converters. Nonetheless, DC microgrids have no standardized voltage, requiring an additional power step to generate AC voltage. Additionally, DC microgrids protection is more complicated.

6.1.2.2 Microgrids control paradigms

The deployment and real-time monitoring of the entire system is made easy by centralized control management. In a centralized control framework, a single individual central controller is the primary controller. In microgrids, the central controller oversees the operation of various microgrid components. Each grid component unit utilizes a local controller to communicate directly with the central controller. With the help of recent computation technologies, the central controller can process and analyze the data streams obtained from the remote local controller in real-time. Implementing centralized control is relatively simple and has shown to be very effective in operating the microgrid system. However, several issues still need to be addressed, especially when dealing with large-scale heterogeneous energy resources. Moreover, central controller failures can have a significant impact on the functionality of the entire system. Additionally, the control technique has limited flexibility and expandability.

Over the past few years, there has been a significant advancement in developing decentralized control techniques for microgrids, which aim to enhance the coordination of energy sources and loads. The primary focus of this approach is to ensure stability, reliability, and cost-effective operation. The local measurements are used to enable global control decisions, which require only a limited number of local connections, and do not mandate high-performance computer units or a high level of connectivity. However, it is important to note that this method does not guarantee global optimum solutions for the entire MG system.

Recently, the need for increasing the hosting capacity of renewable energy sources has led to the development of advanced control and supervision systems for microgrids, which aims at enhancing the power system flexibility and lowering the complexity of microgrid operations. The aim of these new technologies is to enhance the reliability, security, and energy conversion efficiency of microgrids, supporting the massive deployment of renewable power generators, and the mitigation of their grid impacts. Some of the most promising enabling technologies include the following:

- Smart energy management systems: Modern energy management systems employ advanced technologies such as computational intelligence, machine learning, and bio-inspired optimization algorithms to effectively schedule the power generators, the charging/discharging profiles of the energy storage systems, and the flexible loads within microgrids. These systems constantly analyze and predict the energy demand and generation profiles, dynamically dispatch the energy production, and promptly enable reliable decision-making processes aimed at attaining effective operational performance.
- Adaptive tools for energy storage systems management: Advanced tools aimed at controlling energy storage systems have been developed in the task of enhancing their energy conversion efficiency, and their lifetime in microgrids.

These techniques integrates smart scheduling and adaptive modeling techniques, which aim at identifying the optimal charging/discharging profiles on the basis of the real-time capacity demand, renewable power availability, and power grid state. These smart tools enable effective power balancing, peak reduction, load shifting, and grid frequency regulation.

- Grid-forming inverter control: The control of grid-forming inverters has gained attention due to their effectiveness in controlling the voltage and frequency of microgrids independently, thereby allowing the complete independence from the power distribution grid. This feature is extremely relevant in the context of de-carbonized power systems, which will be characterized by a prevalence of renewable generators connected to the grid by power converter-based interfaces. In this context, advanced control algorithms for grid-forming inverters can contribute in increasing the power system stability and enhancing the resilience of microgrids to severe perturbations.
- Demand-side management: Microgrid control systems are now incorporating demand response programs, which allow consumers to actively participate in managing their load. The integration of advanced demand response algorithms and pervasive communication networks facilitate real-time interaction between microgrid operators and end-users, which ultimately optimizes energy consumption by enabling effective load shedding or shifting.
- Cybersecurity tools: As microgrid control systems become increasingly interconnected and reliant on digital communication and control technologies, it is crucial to ensure their cybersecurity and resilience. To mitigate cyber threats, all the computing systems incorporate advanced security features such as encryption, authentication protocols, anomaly detection, and intrusion prevention systems.

6.2 A decentralized framework for smart microgrids operation

This section introduces the foundational concepts behind a decentralized and self-organizing framework designed for synchronized SMG state estimation, monitoring, and control. The core principle draws inspiration from the theory of self-organizing dynamic agents, responsible for executing fundamental SMG functions [1]. Within our conceptualization, a dynamic agent is characterized by the following attributes:

1. It is situated on a specific MG component.
2. It establishes direct connections with locally available sensors/controllers.
3. It communicates with a limited number of neighboring agents.
4. It is equipped with a first-order dynamic system, specifically an oscillator, whose state is initialized by local data sensed by the agent, and interlinked with the states of the nearby oscillators through local coupling strategies.

The forthcoming sections illustrate how this network of dynamic agents can be designed to achieve consensus on global functions from the field data collected by all agents. This biologically inspired approach equips dynamic agents with the ability to solve the main SMG operation problems exclusively by processing locally available information. In particular, the deployment of consensus protocols endows local sensors/controllers with the capability for time synchronization, enabling each dynamic agent to estimate the most pivotal variables defining the holistic SMG operation, without the need for central fusion facilities. Consequently, all fundamental SMG functions, encompassing control, monitoring, and time synchronization, can be executed through an entirely decentralized and non-hierarchical computational framework. This design inherently grants this computing architecture remarkable scalability, flexibility, resilience, and distribution features, rendering it an enabling technology for addressing the complexities of SMGs.

6.3 Time-synchronization of the SMG components

The distributed consensus algorithms detailed in Chapter 2 offer a viable option for achieving synchronization among the dynamic agents clocks, eliminating the necessity for a centralized synchronization infrastructure. Without loss of generality, we assume that the clock τ_i of the ith agent can be represented by a linear equation.

$$\tau_i(t) = \alpha_i\, t + \beta_i \tag{6.1}$$

In this context, the coefficients α_i and β_i, which correspond to the skew and offset parameters, are unknown. To deduce these coefficients, we employ the average time synchronization (ATS) protocol [27]. This protocol falls under the category of distributed algorithms rooted in the average consensus theory. The underlying principle of this bio-inspired paradigm is to gather indirect information concerning the ith dynamic agent unknown clock coefficients. This is accomplished by sharing the time offsets with neighboring agents, ultimately leading to the synchronization of the dynamic agent network with a virtual reference clock:

$$\tau_v(t) = \alpha_v\, t + \beta_v \tag{6.2}$$

Deriving from this definition, it becomes apparent that each dynamic agent possesses the capability to estimate clock parameters through the utilization of a linear regression technique. Indeed, upon synchronization, the dynamic agents ultimately converge asymptotically toward the following shared global reference time:

$$\hat{\tau}_i(t) = \hat{\alpha}_i\, t + \hat{\beta}_i \tag{6.3}$$

where the parameters $\hat{\alpha}_i$ and $\hat{\beta}_i$ are estimated by each individual agent through the application of the aforementioned consensus algorithms [27,28]. Once identified these correction parameters, all the dynamic agent clocks converge to an unique global reference time, namely

$$\lim_{t \to +\infty} \hat{\tau}_i(t) - \tau_v(t) = 0 \;\; \forall i \in [1,N] \tag{6.4}$$

and by observing that

$$\hat{\tau}_i(t) = \hat{\alpha}_i\, \alpha_i\, t + \hat{\alpha}_i\, \beta_i + \hat{\beta}_i \tag{6.5}$$

it follows that

$$\begin{aligned} \lim_{t\to+\infty} \hat{\alpha}_i\, \alpha_i &= \alpha_v \\ \lim_{t\to+\infty} \hat{\alpha}_i\, \beta_i + \hat{\beta}_i &= \beta_v \end{aligned} \tag{6.6}$$

Hence, when the dynamic agents attain consensus, their individual local clocks synchronize, obviating the requirement for a designated cluster header. In this scenario, each agent operates as an autonomous header, transmitting signals to neighboring agents. Consequently, the inherent agent oscillators become synchronized in a unified phase, even in the presence of distinct clock frequencies.

6.4 SMG monitoring

The distributed consensus protocols empower the dynamic agent network to collectively converge on a consensus on the global variables that define the SMG operation. In particular, if the dynamic agents acquire the bus voltage magnitude and the injected active power, the definition of the following observation vector allows the dynamic agents to reach a consensus on several important grid variables:

$$\omega_i = [V_i, N\,(V_i - V^*)^2, (V_i - V^*)^2, N\,P_{Gi}, N\,P_{Li}, N\,(P_{Gi} - P_{Di})] \tag{6.7}$$

where V_i is the ith bus voltage magnitude, V^* is the rated grid voltage, and P_{Gi} and P_{Di} are the active power generated and demanded at the ith bus, respectively. In particular, this observation vector allows the agents network to reach a consensus on the following variables:

$$\omega^* = \left[\frac{1}{N}\sum_{i=1}^{N} V_i, \sum_{i=1}^{N}(V_i - V)^2, \frac{1}{N}\sum_{i=1}^{N}(V_i - V)^2, \sum_{i=1}^{N} P_{Gi}, \sum_{i=1}^{N} P_{Di}, \sum_{i=1}^{N} P_{Gi} - P_{Di}\right] \tag{6.8}$$

which allow the dynamic agents to infer essential SMG metrics, including the average SMG voltage, total voltage magnitude variation, mean voltage fluctuation, the summation of generated and demanded power, and the active power losses. Moreover, by defining alternative observation vectors, additional useful grid variables such as maximum and minimum voltage angles, as well as power quality indices, can be effortlessly computed. This pivotal attribute empowers every agent to autonomously identify local grid irregularities, enabling the system operator to gain insight into the comprehensive state of the global SMG operation by consulting any agent within the network.

6.5 Decentralized estimation and control

By implementing the monitoring functions delineated in Section 6.4, each dynamic agent gains the ability to evaluate both the local performance of the monitored bus and the overarching performance of the entire SMG. This setup allows for a continuous comparison of local and global performance indices, rendering it possible to promptly respond when a substantial deviation arises between the local performance index and its corresponding system-wide counterpart. This adaptive response mechanism can be harnessed by the network of dynamic agents for the purpose of decentralized SMG control.

Bolstered by this strategic perspective, SMG control paradigms have been devised to enable the utilization of decentralized consensus protocols in orchestrating the behavior of responsive dynamic agents. These intelligent entities can assume control decisions grounded in the interplay between local and global SMG conditions. The focal point of these control paradigms revolves around addressing two pivotal SMG quandaries: voltage control and power flow analysis.

The mathematical backbone supporting these functions is the capability of the dynamic agents network in solving this fundamental problem in a decentralized way:

$$y = \begin{bmatrix} y_1 \\ \dots \\ y_m \end{bmatrix} = A\omega = \begin{bmatrix} A_{11} & \dots & A_{1N} \\ \dots & \dots & \dots \\ A_{m1} & \dots & A_{mN} \end{bmatrix} \begin{bmatrix} \omega_1 \\ \dots \\ \omega_N \end{bmatrix} \tag{6.9}$$

where y denotes an unknown vector, A represents a known coupling matrix, and ω is the vector of the local observations. In our studies, we suppose that the ith agent knows ω_i and the ith column of the coupling matrix (namely, $[A_{1i}...A_{Ni}]^T$). Hence, the decentralized solution of this problem can be obtained by deploying a cooperative scheme based on distributed average consensus protocols. For this purpose, the following observation vector is defined for the ith dynamic agent:

$$\pi_i = \begin{bmatrix} N\,A_{1i}\,\omega_i \\ \dots \\ N\,A_{Mi}\,\omega_i \end{bmatrix} \tag{6.10}$$

This choice allows the dynamic agents network to synchronize to the following vector:

$$\omega^* = \begin{bmatrix} \sum_{i=1}^{N} A_{1i}\,\omega_i \\ \dots \\ \sum_{i=1}^{N} A_{Ni}\,\omega_i \end{bmatrix} \tag{6.11}$$

which allows each dynamic agent to compute all the components of the unknown vector y in an entirely decentralized way.

These consensus protocols empower the network of dynamic agents to successfully solve challenging issues, encompassing decentralized state estimation, hypothesis testing, collective optimization, and decision-making. Within the context of signal processing literature, these decentralized problem-solving capacities are commonly referred as “in-field computing.”

6.5.1 Voltage control

The primary objective of SMG voltage control lies in the determination of the optimal set points of the local (primary) controllers, which allows for the enhancement of both the SMG voltage profile and the reduction of power losses. In the envisaged control paradigm, it is presupposed that each local voltage controller operates under the supervision of a dynamic agent. These agents are entrusted with the responsibility of governing the reactive power flow injected into the SMG.

To effectively implement this control strategy, the dynamic agents should initially undertake the computation of global variables that encapsulate the actual operation state of the SMG, as outlined in Section 6.4. Subsequently, the average SMG voltage magnitude is categorized into distinct classifications: "Normal," "Under-Voltage," or "Over-Voltage" [1]. Armed with this classification, the ensuing control scheme amalgamates local and global information implementing the following control rules:

1. If the SMG voltage magnitude is categorized as "Under-Voltage" ("Over-Voltage"), the dynamic agents strive to increase (decrease) the mean SMG voltage magnitude V_M by adjusting the injected reactive power flows using a proportional control law, namely $\alpha|V_M - V_i|$.
2. In the case of a "Normal" SMG voltage magnitude, the local controllers focus on diminishing the voltage magnitude deviation. They achieve this by adjusting the injected reactive power flows upwards (downwards) when $V_M > V_i$ ($V_M < V_i$).

More sophisticated control paradigms that endeavor to address the challenge of SMG voltage control by employing advanced optimization techniques (such as collective optimization or cooperative heuristic optimizers) can be aptly employed through the dynamic agents network. However, it is important to note that the adoption of these optimization-driven approaches might necessitate significantly higher computational resources and communication demands when compared to a centralized approach. Additionally, considering the expected operation of SMGs in constrained operational scenarios, the likelihood of encountering infeasibility due to communication issues is considerably amplified.

6.5.2 *Power flow analysis*

Power flow analysis constitutes a fundamental element for numerous SMG management tools, encompassing network optimization, state estimation, and service restoration [3]. Its primary objective is to compute the operating state of the SMG under fixed load demand and power generation conditions. Conventionally, the algorithm utilized to solve this issue in radial or moderately meshed SMG configurations relies on a pair of matrices that define the SMG's topology, namely [29]:

1. The bus-injection to branch current matrix (A^{BIBC}), which allows modeling the relationship between the N bus current injections $A^{BIBC} = [J_1, ..., J_N]^T$ and the NL branch currents $B = [B_1, ..., B_{NL}]^T$ as follows:

$$B = A^{BIBC} A^{BIBC} \tag{6.12}$$

2. The branch current to bus voltage matrix (A^{BCBV}), which allows modeling the relationship between the NL branch currents and the N bus voltages as follows:

$$\Delta V = A^{BCBV} B \tag{6.13}$$

The definition of these two matrices allows determining the SMG bus voltage phasors by solving the following set of equations:

$$\Delta V = A^{BCBV} A^{BIBC} J = A^{DLF} J \tag{6.14}$$

Consequently, the dynamic agents network can solve the SMG power flow problem by using the following decentralized solution scheme:

1. define an initial power flow solution $V^{(0)} = [V_1^{(0)}, ..., V_N^{(0)}]^T$;
2. by using the current solution $V_i^{(k)}$ each dynamic agent determines its bus current injection J_i;
3. solve an average consensus problem by considering the following observation vector:

$$\pi_i = \begin{bmatrix} N\,A1_i^{DLF}\ V_i^{(k)} \\ \dots \\ N\,AN_i^{DLF}\ V_i^{(k)} \end{bmatrix} \tag{6.15}$$

   where each agent determines the elements of the matrix A^{DLF} by using the algorithm described in [29]. This decentralized protocol allows the dynamic agents to reach a consensus on the following state:

$$\omega^* = \begin{bmatrix} \sum_{i=1}^{N} A1_i^{DLF}\ V_i^{(k)} \\ \dots \\ \sum_{i=1}^{N} AN_i^{DLF}\ V_i^{(k)} \end{bmatrix} = \Delta V \tag{6.16}$$

4. update the bus voltage magnitudes:

$$V^{(k+1)} = V^{(k)} + \Delta V \tag{6.17}$$

5. terminate if a proper convergence criterion is satisfied, else go to step 3.

6.6 Example

This section analyzes the performance of the described decentralized framework applied to the SMG depicted in Figure 6.1 [3].

Within this electrical network context, it was assumed the presence of a voltage-following wind generator (i.e., equipped with an induction generator directly linked to the grid), and two voltage-supporting wind generators (i.e., linked to the grid via a power electronic interface). Key parameters characterizing these distributed generating systems have been presented in Table 6.6 for reference.

Regarding the voltage regulators, our focus encompasses a line tap-changing transformer situated at Bus 1, along with a shunt compensator with a designated nominal power of 2 MVA situated at bus 9.

To simulate the SMG, a network comprising of 18 dynamically linked agents was implemented. The coefficients of the adjacency matrix governing the communication network topology were determined based on the logical boundaries outlined by the dashed circles as illustrated in Figure 6.1. Communication delays were assumed to be constant but unspecified for each agent.

The complete range of the described decentralized functions was effectively employed in the context of this test system. To assess the self-organizing

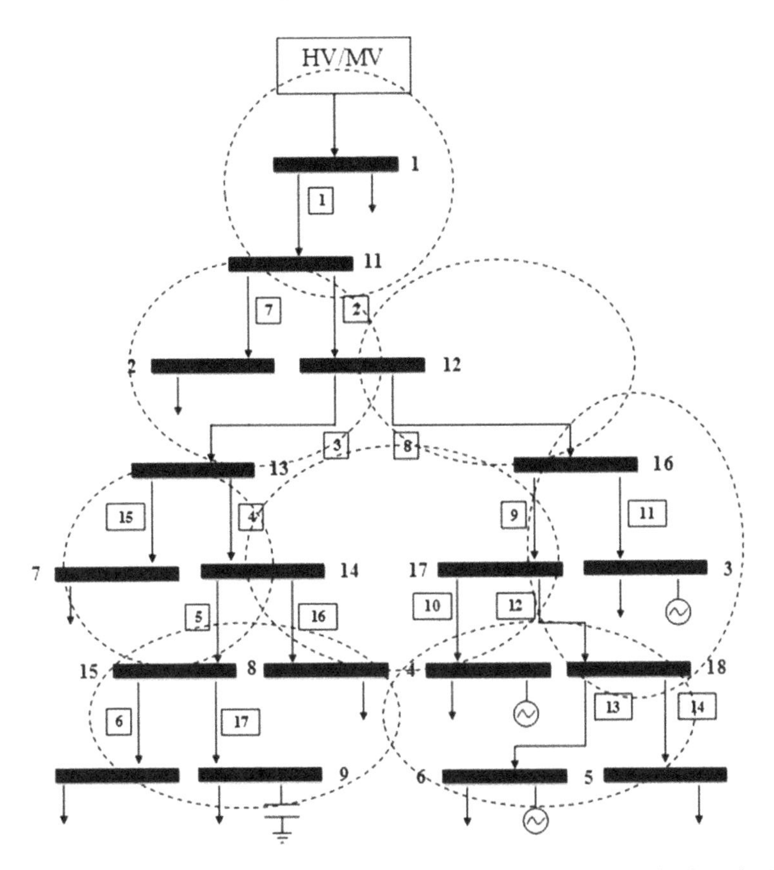

Figure 6.1 Analyzed SMG and dynamic agents boundary (dashed circles)

Table 6.1 Main characteristics of the local generators

Operation mode	**Power profile**	**Bus**	Q_{min} **(MVar)**	Q_{max} **(MVar)**	P_{nom} **(MVA)**
Voltage supporting	1	3	–0.6	0.72	1.2
Voltage supporting	2	4	–0.6	0.72	1.2
Voltage following	3	6	–	–	1.2

architecture effectiveness in achieving synchronized sensor time-frames, we introduced a scenario where the network encompassed 18 unsynchronized local clocks. The coefficients of these clocks were assumed to uniformly vary within the specified ranges:

$$\begin{aligned} \alpha_i &\in [0,1] \\ \beta_i &\in [0,1]\ \forall i \in [1,18] \end{aligned} \tag{6.18}$$

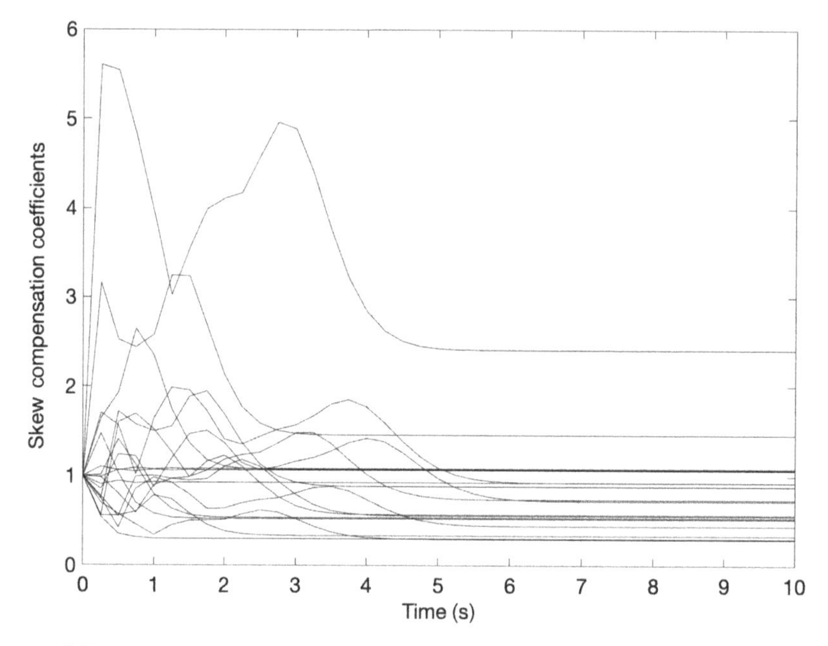

Figure 6.2 Time synchronization function: agents state trajectories in the task of computing the skew correction coefficient $\hat{\alpha}_i$

Starting from these initial values, the implementation of the decentralized consensus protocol for time synchronization yields the results summarized in Figures 6.2–6.4. In particular, the trajectories of the dynamic agents states, during their endeavor to compute skew and offset correction coefficients, are presented in Figures 6.2 and 6.3, respectively. Moreover, Figure 6.4 illustrates the corresponding progression of the synchronized local clocks. Analyzing these profiles unveils that, following an initial transient adjustment, the clocks converge harmoniously to a singular virtual reference point. This underscores the effectiveness of the decentralized framework in time-synchronizing all the measurements, a feature that can be particularly advantageous in scenarios where the primary SMG synchronization source encounters disruption or degradation.

Further simulation investigations were undertaken to assess the efficacy of the dynamic agents network in the context of SMG power flow analysis. In this scenario, the resident dynamic agent at bus 1 gauged the corresponding bus voltage magnitude, while the remaining agents gauged the active and reactive power demands and generation at the monitored bus. The outcomes of these studies are encapsulated in Figure 6.5, which reports the agents state trajectories in calculating the bus voltage magnitude for a specific operational state of the SMG. An insightful observation from this figure reveals that the agents network adeptly solves the power flow problem within two main iterations.

To substantiate the resilience of the decentralized approach, we solved the SMG power flow problem in the presence of varying load and generation profiles, which were characterized by the time profiles reported in Figures 6.9 and 6.10, respectively.

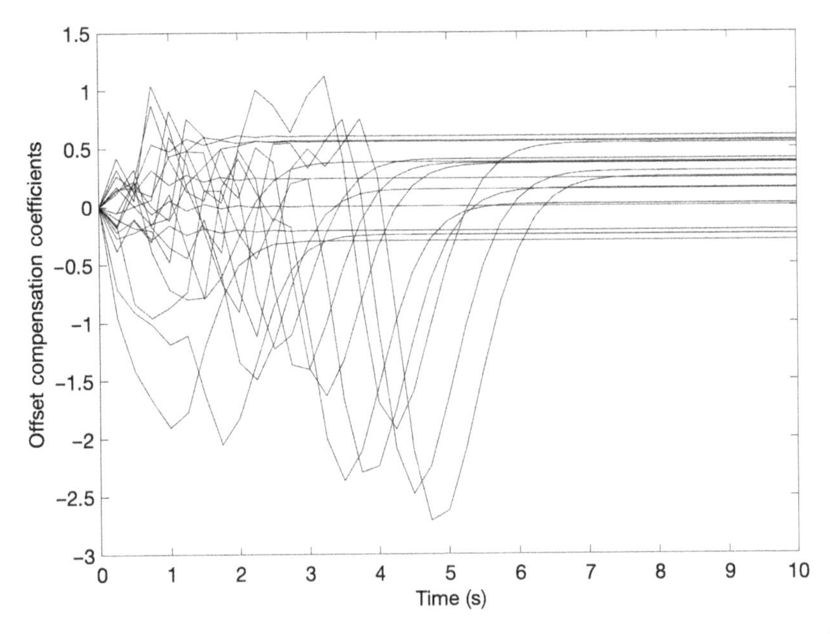

Figure 6.3 Time synchronization function: agents state trajectories in the task of computing the offset correction coefficient $\hat{\beta}_i$

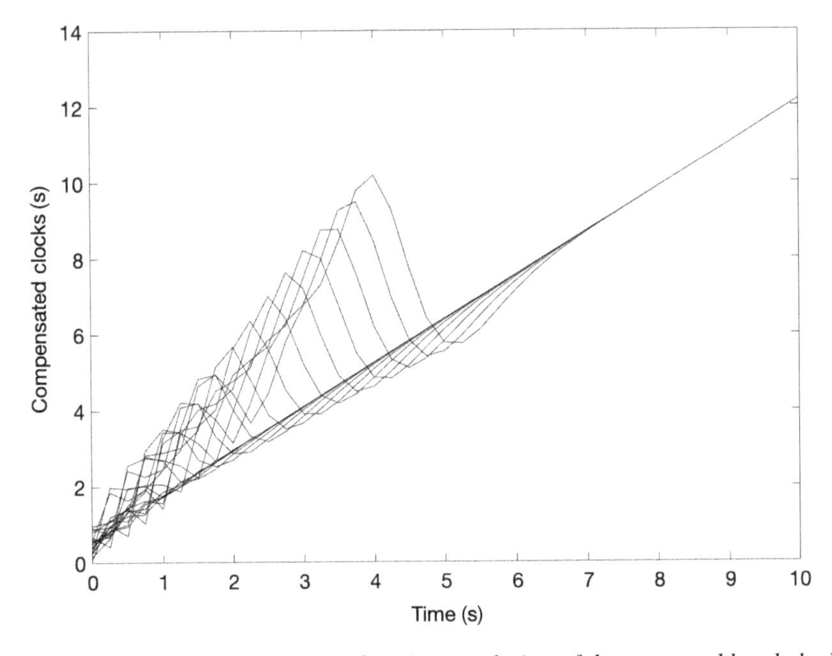

Figure 6.4 Time synchronization function: evolution of the corrected local clocks

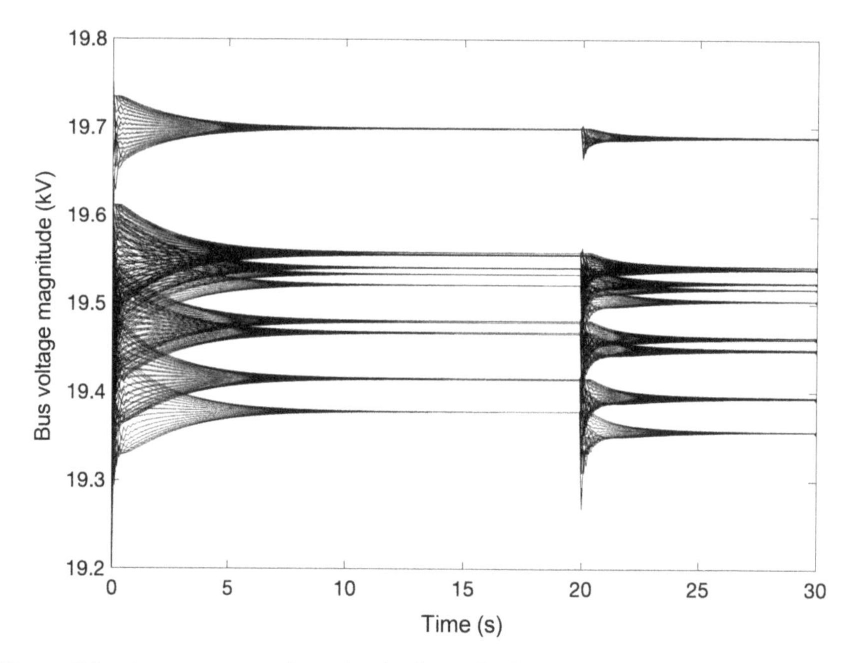

Figure 6.5 Agents state trajectories in the task of computing the SMG bus voltage magnitudes

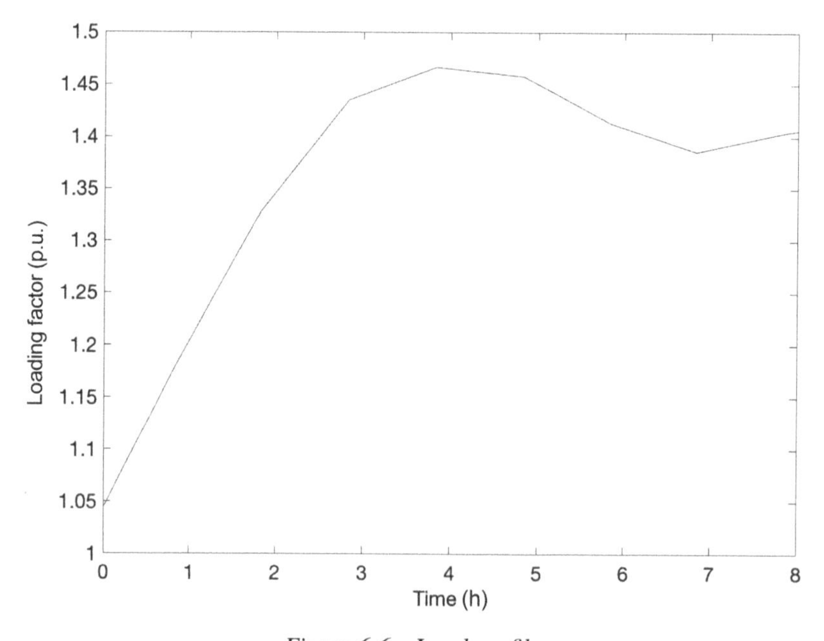

Figure 6.6 Load profile

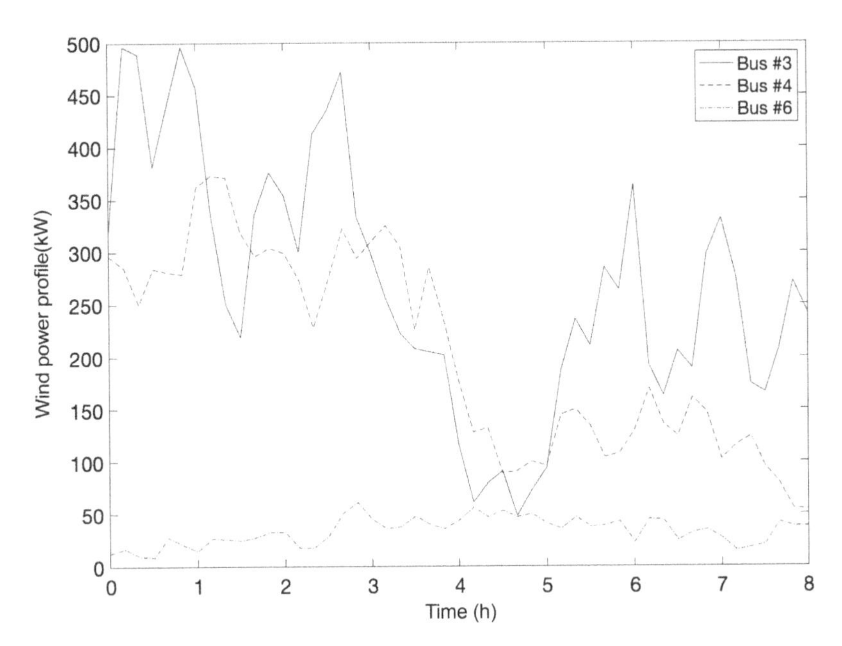

Figure 6.7 Wind power profile

The solutions of the power flow problem in the presence of these highly variable profiles computed by a centralized solution algorithm are reported in Figure 6.8, while the corresponding solutions computed by the dynamic agents, in terms of bus voltage magnitude and line current magnitude, are reported in Figures 6.9 and 6.10, respectively. By analyzing these results, it is worth observing as the solutions computed by the decentralized solution scheme accurately track the variable SMG operation state.

The derived power flow solutions can be effectively leveraged to underpin various SMG control functions. Particularly, these solutions can be adopted to solve the SMG voltage control problem in accordance with the decentralized control scheme outlined in Section 6.5.1. The dynamic agents stationed at buses 1, 3, 4, 6, and 9 take charge of SMG voltage-regulating devices, thus implementing control actions. The corresponding trajectory of agent states implementing the SMG voltage control are depicted in Figure 6.11. Within each control iteration, dynamic agents determine the bus voltage magnitude profile through power flow analysis, regulating the generated reactive power flows. Subsequently, agents reevaluate the SMG operation state repercussions due to these control actions through a new power flow analysis. This voltage control loop terminates when a predetermined termination criterion is met.

The impacts of the decentralized voltage control strategy on SMG voltage enhancement can be gauged through an examination of Figure 6.12. An analysis of this figure shows that dynamic agents effectively address the voltage regulation problem within approximately 10 control iterations (equivalent to 4600 time steps). This

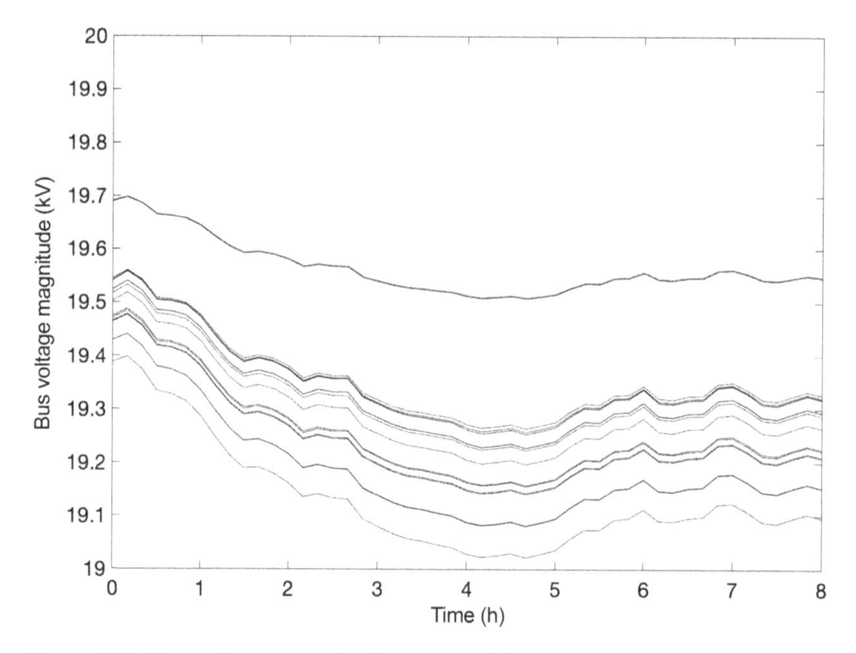

Figure 6.8 Bus voltage magnitude computed by a centralized solution scheme

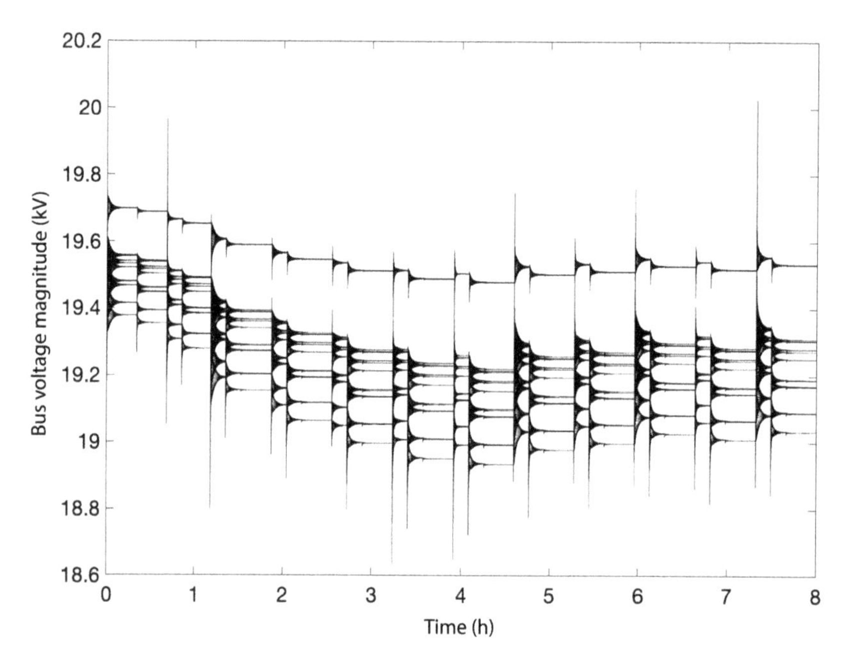

Figure 6.9 Bus voltage magnitude computed by the decentralized framework

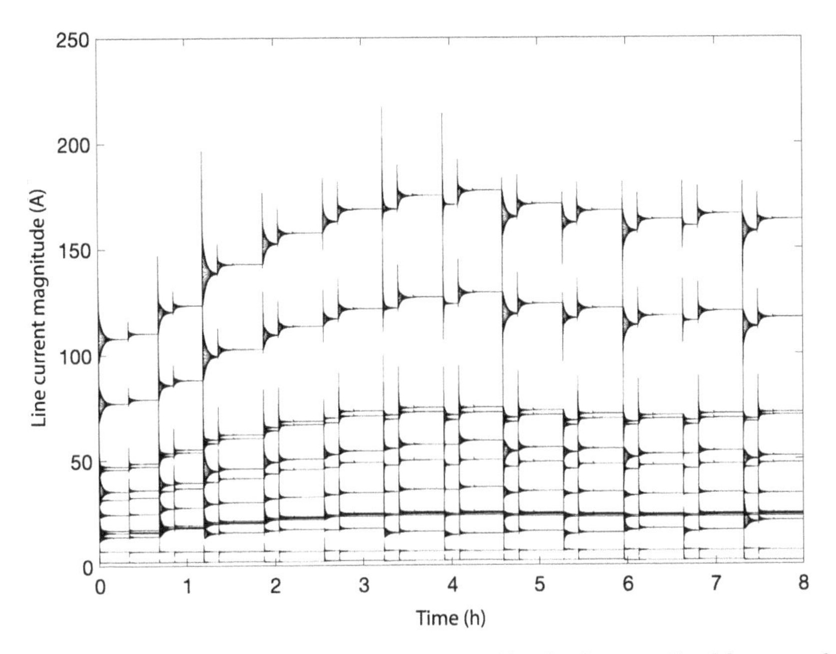

Figure 6.10 Line current magnitude computed by the decentralized framework

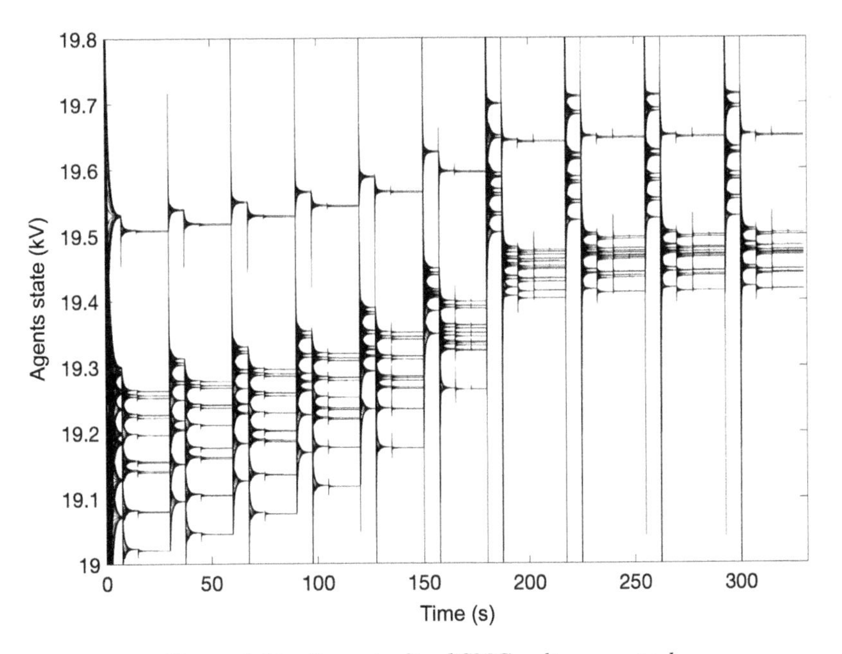

Figure 6.11 Decentralized SMG voltage control

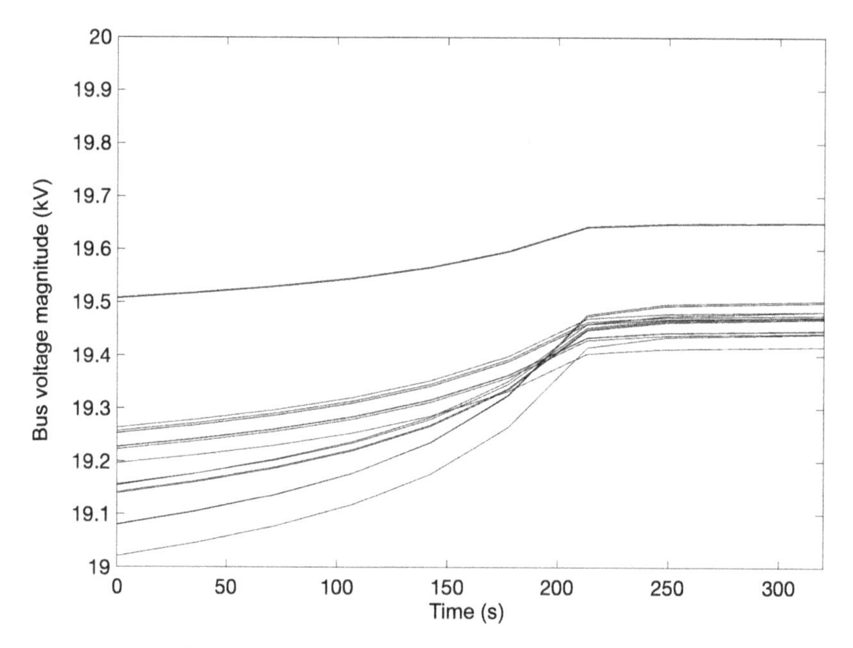

Figure 6.12 Grid impacts of the decentralized voltage control

point highlights that these accomplishments were realized within a non-hierarchical and decentralized paradigm. Notably, all the requisite data for executing SMG functions were generated by dynamic agents without necessitating centralized processing or storage infrastructure.

Lastly, in the context of the practical implementation of the proposed framework, it is worth underscoring the following considerations:

1. The computational load primarily arises from solving a set of mutually interdependent first-order differential equations. The corresponding rate of convergence is determined by the eigenvalues of the adjacency matrix. This insight facilitates a theoretical assessment of how communication delays and fluctuations in link availability might impact the algorithm convergence and stability. Furthermore, the time required for achieving consensus within linearly interconnected networks is exclusively influenced by specific graph properties characterizing the agents' logical interconnections.
2. A comprehensive demonstration found in [30] underscores that the decentralized consensus protocols will invariably converge, even in communication networks characterized by dynamically evolving topologies, under the precondition that infinitely recurring communication graphs maintain interconnectedness. This pivotal attribute fortifies the network against the inclusion or departure of agents, and network synchronization can be reliably maintained, exhibiting distinct convergence rates for varying agent numbers, achieved through proper communication graph design.

3. The described decentralized framework attains convergence to an average consensus state, even when confronted with instances of random communication link disruptions, as confirmed by [31].

These attributes could significantly enhance the seamless and effective implementation of this computing architecture within modern SMG. In light of this, it becomes important to develop novel semantic tools tailored to the acquisition and processing of operational data in standardized and interoperable formats. Consequently, many research endeavors have been directed toward enhancing self-organizing framework by integrating specialized tools capable of harnessing the semantic representation of measurements within SMGs. Anticipated benefits encompass augmented decision support for SMG operators, especially if the focus remains on enabling dynamic agents to engage at a conceptual level familiar to these human operators. Additionally, this technology empowers SMG operators to model the semantic aspects of data, transcending mere reliance on syntactic and structural representations.

References

[1] Vaccaro A, Loia V, Formato G, *et al.* A self-organizing architecture for decentralized smart microgrids synchronization, control, and monitoring. *IEEE Transactions on Industrial Informatics*. 2015;11(1):289–298.

[2] Saeed MH, Fangzong W, Kalwar BA, *et al.* A review on microgrids' challenges and perspectives. *IEEE Access*. 2021;9:166502–166517.

[3] Vaccaro A, Popov M, Villacci D, *et al.* An integrated framework for smart microgrids modeling, monitoring, control, communication, and verification. *Proceedings of the IEEE*. 2011;99(1):119–132.

[4] Chalah S, Belaidi H, Merrad L, *et al.* Microgrid energy management strategy based on MAS. In: *2022 3rd International Conference on Human-Centric Smart Environments for Health and Well-Being (IHSH)*; 2022. p. 1–6.

[5] Tsikalakis AG and Hatziargyriou ND. Centralized control for optimizing microgrids operation. In: *2011 IEEE Power and Energy Society General Meeting*; 2011. p. 1–8.

[6] Ghorbanian M, Dolatabadi SH, Masjedi M, *et al.* Communication in smart grids: a comprehensive review on the existing and future communication and information infrastructures. *IEEE Systems Journal*. 2019;13(4):4001–4014.

[7] Williams B, Gahagan M, Dromey I, *et al.* Using distributed decision-making to optimize power distribution and support microgrids. In: *2012 IEEE Power and Energy Society General Meeting*; 2012. p. 1–6.

[8] Phillips LR. Managing microgrids using grid services. In: *2007 IEEE International Conference on System of Systems Engineering*; 2007. p. 1–5.

[9] De Brabandere K, Vanthournout K, Driesen J, *et al.* Control of microgrids. In: *2007 IEEE Power Engineering Society General Meeting*; 2007. p. 1–7.

[10] Colson CM, Nehrir MH, and Gunderson RW. Distributed multi-agent microgrids: a decentralized approach to resilient power system self-healing. In:

2011 4th International Symposium on Resilient Control Systems; 2011. p. 83–88.
[11] Pepiciello A. Flexibility solutions for the integration of variable renewable energy sources in power systems. PhD Thesis, University of Sannio; 2021.
[12] Zhang G, Jiang C, and Wang X. Comprehensive review on structure and operation of virtual power plant in electrical system. *IET Generation, Transmission & Distribution*. 2019;13(2):145–156.
[13] Mashhour E and Moghaddas Tafreshi SM. Bidding strategy of virtual power plant for participating in energy and spinning reserve markets, part I: problem formulation. *IEEE Transactions on Power Systems*. 2010;26(2):949–956.
[14] Yang H, Yi D, Zhao J, *et al.* Distributed optimal dispatch of virtual power plant via limited communication. *IEEE Transactions on Power Systems*. 2013;28(3):3511–3512.
[15] Morales JM, Conejo AJ, Madsen H, *et al. Integrating Renewables in Electricity Markets: Operational Problems*. vol. 205. New York, NY: Springer Science & Business Media; 2013.
[16] Pepiciello A, Bernardo G, D'Argenzio E, *et al.* A decision support system for the strategic operation of virtual power plants in electricity markets. In: *2019 International Conference on Clean Electrical Power (ICCEP)*. Piscataway, NJ: IEEE; 2019. p. 370–374.
[17] Mancarella P. MES (multi-energy systems): an overview of concepts and evaluation models. *Energy*. 2014;65:1–17.
[18] Geidl M, Koeppel G, Favre-Perrod P, *et al.* Energy hubs for the future. *IEEE Power and Energy Magazine*. 2006;5(1):24–30.
[19] Mancarella P. MES (multi-energy systems): an overview of concepts and evaluation models. *Energy*. 2014;65:1–17.
[20] Krause T, Andersson G, Frohlich K, *et al.* Multiple-energy carriers: modeling of production, delivery, and consumption. *Proceedings of the IEEE*. 2010;99(1):15–27.
[21] Qi F, Wen F, Liu X, *et al.* A residential energy hub model with a concentrating solar power plant and electric vehicles. *Energies*. 2017;10(8):1159.
[22] Almassalkhi M and Hiskens I. Optimization framework for the analysis of large-scale networks of energy hubs. In: *Power Systems Computation Conference*, vol. 1; 2011.
[23] Ha T, Zhang Y, Hao J, *et al.* Energy hub's structural and operational optimization for minimal energy usage costs in energy systems. *Energies*. 2018;11(4):707.
[24] Vahid-Pakdel M, Nojavan S, Mohammadi-Ivatloo B, *et al.* Stochastic optimization of energy hub operation with consideration of thermal energy market and demand response. *Energy Conversion and Management*. 2017;145:117–128.
[25] Fouskakis D and Draper D. Stochastic optimization: a review. *International Statistical Review*. 2002;70(3):315–349.
[26] Uddin M, Mo H, Dong D, *et al.* Microgrids: a review, outstanding issues and future trends. *Energy Strategy Reviews*. 2023;49:101127.

[27] Zhu L, Shi S, and Gu X. A consensus-based distributed clock synchronization for wireless sensor network. In: *2018 14th International Wireless Communications & Mobile Computing Conference (IWCMC)*; 2018. p. 828–832.

[28] Schenato L and Fiorentin F. Average TimeSynch: a consensus-based protocol for clock synchronization in wireless sensor networks. *Automatica*. 2011;47(9):1878–1886.

[29] Teng JH. A direct approach for distribution system load flow solutions. *IEEE Transactions on Power Delivery*. 2003;18(3):882–887.

[30] Patterson S, Bamieh B, and El Abbadi A. Convergence rates of distributed average consensus with stochastic link failures. *IEEE Transactions on Automatic Control*. 2010;55(4):880–892.

[31] Xiao L, Boyd S, and Lall S. A scheme for robust distributed sensor fusion based on average consensus. In: *IPSN 2005. Fourth International Symposium on Information Processing in Sensor Networks, 2005*; 2005. p. 63–70.

Chapter 7

A decentralized framework for dynamic thermal rating assessment of overhead lines

Over recent years, there has been a noticeable rise in electricity demand, a heightened intensity in energy trading activities, and an exponential growth of renewable power generators into existing power grids. These factors are collectively driving the imperative to enhance the transmission and distribution assets by increasing their transmission capacity. Anyway, many nations are grappling with public resistance to the construction of new power lines, driven by concerns over their visual impact and potential environmental impacts.

In this context, increasing the thermal rating for existing transmission and distribution lines emerges as a strategic alternative to the construction of new grid assets. The transmission capacity of an overhead line is essentially determined by the maximum allowable current, or "ampacity," which corresponds to the maximum tolerable temperature of the conductors. This temperature limitation is influenced by both the thermal characteristics of the conductor material and the available phase-to-ground clearances along the span of the line. The temperature of the conductors, in turn, hinges on the line current and the weather conditions along the entire line route [1].

Power line ampacity is restricted by the thermal stress that impacts the mechanical components of the line, leading to an adverse effect on the line lifespan. Moreover, the risk of conductor elongation causing an increased fault risk also limits the capability of transmission lines. Therefore, system operators set a maximum temperature value for the conductor operation and keep the lines below a maximum loading current. This value is calculated based on worst-case conditions for the variables involved in the conductors heat exchange process. However, these worst-case scenarios are highly conservative and seldom occur during the conductor operating life. Therefore, dynamic ampacity assessment based on ambient conditions is necessary to exploit the full power line load capability.

Assessing the dynamic ampacity of a power line entails understanding its ability to sustain a given current while adhering to prescribed ground clearances or conversely, determining the maximum permissible current for a given duration. The degree of dynamic overloading that a power line can accommodate is typically constrained by the conductor temperature. Indeed, for fixed installation conditions, the sag and clearance parameters are heavily influenced by the current and expected conductor temperatures.

For this purpose, real-time modeling of the conductor temperature is required, and the scientific literature presents several approaches to solve this issue. These approaches are classified into two categories, namely direct methods and indirect methods. Direct methods are based on the direct measurement of temperature or directly related quantities, such as sag variation and mechanical stress. On the other hand, indirect methods are based on the estimation of conductor temperature through an energy balance approach that utilizes real-time measurement of the ambient variables in its surroundings, as described in [2].

The concept of dynamic thermal rating (DTR) encompasses the continuous monitoring of conductor temperatures, or the utilization of weather and load forecasts to infer their evolution, and the use of modeling techniques enabling the calculation of the actual transmission line capacity [3–5]. This dynamic estimation approach represents a significant advancement over the conventional static thermal rating (STR) method, which predominantly relies on assuming the most adverse weather conditions within a specified period [6]. Moreover, static thermal ratings neglect the large thermal time constants of overhead lines, which can exceed 10 min, unlocking noteworthy additional transient capacity. By fully harnessing the thermal dynamic behaviors of conductors corresponding to the actual weather conditions, DTR empowers transmission system operators (TSOs) with enhanced dispatching flexibility and enabling reliable decision-making during grid congestions, and enhancing the effectiveness of the corrective actions [7]. Consequently, DTR is progressively being integrated into many important real-time operation tools [8].

In weather-dependent DTR procedures, both the present and future conductor temperatures are estimated using a detailed thermal model of the transmission line [9]. These estimates hinge on assuming that the forthcoming trends in power flow, ambient temperature, solar radiation, and wind speed and direction over the next 2–3 h, as far as their spatial variations are predetermined. Weather inputs, which pose a significant challenge in prediction for TSOs, are typically provided through rolling-horizon forecasting tools, and a network of meteorological stations. By using these data, the tensions, sags, and clearances at each span of the line can be computed at each time interval. However, it remains evident that the precision of local weather forecasts constitutes a weak link in projecting the ampacity of a transmission line for the upcoming few hours, given that load flow predictions are relatively more foreseeable for TSOs [10].

To address this concern, more advanced prediction methodologies for thermal rating, founded on distributed temperature sensing systems, have been proposed in the literature [11]. While laboratory results in these studies showcase the potential of such computing frameworks for dynamic loadability assessment of overhead lines, practical implementation along the line route presents several technical and economic challenges. An encouraging approach to surmount these limitations lies in leveraging cooperative and pervasive sensor networks distributed across the line route [12]. These distributed sensors directly measure conductor temperatures and transmit the results to a central server through short-range communication. This approach empowers TSOs to attain a reliable assessment of the actual line thermal state and precise insight into the location and magnitude of hotspot conductor temperatures. Despite these benefits, the widespread integration of these technologies within modern transmission systems remains in its infancy, and several unsolved issues should be fixed. In this context, designers of sensor network-based DTR prediction systems have defined the technical prerequisites and the design principles that should be considered in developing new DTR technologies [4]:

1. High precision thermal modeling, which ensures accurate analysis of the heat exchange dynamics between the conductors and the external environment.
2. Distributed estimation and detection capabilities, which enable the reliable assessment of spatial profiles for both conductor temperature and environmental variables along the line route.
3. Self-organizing, which allows designing detection and response mechanisms for system faults or data outliers that could compromise the computing accuracy.
4. Global time synchronization, which allows defining a unified time reference for measuring the line temperature and the environmental variables.
5. Adaptive features, which allow modelling the complex, time-varying phenomena influencing the conductor thermal behavior.
6. High scalability, which allows developing flexible and cost-effective DTR systems for multiple line configurations.

To address these challenges, this chapter analyzed the role of decentralized sensors network equipped with the consensus protocols described in Chapter 2 for DTR assessment of overhead power lines. The main idea is to deploy the smart sensors along the line route, which:

1. measure local weather variables through a set of meteorological sensors;
2. compute the conductor temperatures through a calibrated thermal model;
3. validate the computed temperatures;
4. share their computations through decentralized consensus protocols.

This DTR framework avoids the need for a centralized server while enabling the time-synchronization of the local measurements, and the accurate conductor temperature predictions. Moreover, each smart sensor can compare local conductor temperatures with estimated maximum, minimum, and average values, allowing for real-time comparisons between local and global variables and implementing corrective actions if sensor measurements significantly deviate from predetermined confidence intervals. Experimental results obtained from a real case study will be presented and discussed in order to demonstrate the effectiveness of the self-organizing DTR framework in realistic operation scenarios [4].

7.1 Related works

Calculating the DTR of overhead power lines requires a complex prediction process that involves identifying the location of the hottest temperature spot and forecasting its time evolution for different hypothetical load levels. The process takes into account the actual thermal state of the conductor and the forecasted environmental conditions. This spatial and time-based prediction process is challenging because it requires identifying the critical span, where the heat exchange parameters are at their worst, and predicting its time evolution for time horizons ranging from a few minutes to several hours. To address this challenge, in [13] a cost-minimization approach is proposed for selecting the critical spans, where the most important meteorological variables that affect the heat exchange phenomena between the line conductor and the environment, such as wind speed and direction, are acquired and processed.

A recent solution to this issue involves using a meta-heuristic optimization procedure that utilizes historical-simulated weather data and statistical analysis of thermal capacities computed for each span along the line. This approach helps in identifying the critical spans where the worst-case heat exchange parameters are expected, and predicting their time evolution for different hypothetical load levels. By using historical weather data and statistical analysis, this approach can help in making more accurate predictions and improving the DTR of overhead power lines [14].

The experimental results obtained by applying these methods have led to the development of several standard procedures for indirectly estimating the ampacity of overhead power lines. These procedures are based on detailed conductor thermal models and utilize the insights gained from experimental data analysis [15]. The importance of this topic is further highlighted by the recent publication of a technical brochure by the Cigrè committee [15,16], which provides guidelines and recommendations for the calculation of the thermal rating of overhead power lines and emphasizes the need for accurate estimation of the conductor temperature to ensure the safe and reliable operation of the power grid.

Several technical limitations of the ordinary calculation procedures have been outlined in [2]. The latter mainly derived from the strong uncertainties affecting the thermal model input parameters, essentially related to the radiated heat loss, the solar heat absorbed, the conductor surface properties, the level of atmospheric pollution, the conductor aging, and so on. More specifically, as highlighted in [17], the conductor heat loss radiated in the surrounding atmosphere depends on the surface condition through a random coefficient of emissivity.

According to [2], conventional calculation methods have a number of technical limitations, which are largely attributed to the uncertainty characterizing thermal model input parameters. Some of the factors that contribute to this uncertainty include the randomness of heat loss due to radiation, the amount of solar heat absorbed, the conductor surface properties, the level of pollution in the atmosphere, and the aging of the conductor. In particular, the randomness of heat loss radiated by the conductor into the surrounding atmosphere is dependent on the surface condition and can be characterized by a random coefficient of emissivity [17]. Moreover, the amount of solar heat absorbed by a conductor is also affected by surface conditions and pollutant levels, which can lead to significant variations in the absorption coefficient. This is another source of uncertainty that impacts the accuracy of thermal modeling. Furthermore, the values of meteorological variables used in thermal rating assessment procedures can also introduce additional uncertainties in the thermal model input parameters. Therefore, it is important to carefully consider and account for all these sources of uncertainty when developing and using thermal models for load capability assessment.

Finally, the analyst needs to take into account that weather data obtained from weather providers or local weather stations in real-time may exhibit significant spatial variations. This implies that observations made at one location may not accurately reflect the average conditions along the entire line route. Therefore, the thermal rating of each span will differ, even if measured at the same point in time, due to the weather conditions fluctuations along the line route [18].

In this context, experimental studies demonstrated that the uncertainties affecting environmental temperature and solar radiation can be reliably modeled, while the most challenging issue to address is modeling the spatial variation and the time randomness of both the wind speed and the direction.

These findings suggest that more accurate and reliable calculation methods are needed to account for the complexities and uncertainties associated with thermal modeling [10].

In recent years, the widespread installation of phasor measurement units (PMUs) in many power systems has led to the development of a new approach for estimating the temperature of transmission line conductors [6]. This approach involves processing synchronized phasor data measured at both ends of a transmission line to estimate temperature based on the correlation existing between temperature and direct sequence electric parameters of the line, specifically the line electric resistance.

One of the key advantages of this method is that it eliminates the need to deploy DTR-dedicated sensors and telecommunication infrastructure, as it requires only data already available to system operator's control centers for other purposes [19]. The electric parameters that characterize the line, such as resistance, inductance, and capacitance, can be used to estimate the line temperatures based on their dependency on conductor operating temperature [19].

Several methodologies have been developed to estimate these electric parameters, ranging from simple single or double measurement methods to more sophisticated techniques based on linear and non-linear least square procedures, non-linear programming optimization, or a combination of these techniques. However, uncertainties in data and model parameters can affect the accuracy of temperature estimation in real operating scenarios [19].

A study reported that the application of these methodologies in real operating scenarios produced low-accuracy temperature estimations due to the impact of uncertainty in the measured data [2]. Therefore, further research is needed to develop more refined techniques that can overcome these issues and improve the accuracy of temperature estimations.

With the widespread use of modern distributed temperature sensing modules based on fiber optical technology, advanced techniques for DTR have been introduced in power systems literature [20]. These techniques integrate dynamic thermal models with computational intelligence-based methods to process measured data obtained using distributed fiber-optic sensors, which can accurately measure the conductor temperature profile along the line route. The solution of the dynamical thermal models allows:

1. estimating the actual set of thermal parameters model (i.e. thermal parameters calibration),
2. identifying the line span that is characterized by the worst-case heat exchange conditions (i.e. the critical line span),
3. computing the highest conductor temperature along the line route (i.e. hotspot temperature),
4. assessing the dynamic ampacity of the power line.

The findings presented in the literature suggest that distributed temperature sensing modules that operate on fiber optics can be advantageous for predicting DTR of overhead power lines. However, the installation of optical fibers throughout the line route and their connection to the line conductor can present various technical and financial hurdles. To address these issues, one promising research option is to utilize cooperative and pervasive sensor networks that are distributed along the line route.

The use of distributed sensor networks for real-time assessment of overhead line loadability has been proposed in a number of research papers, including [21]. The authors of these papers have employed various techniques, such as Multi-Layer Perceptron Neural Network and Echo State Network, to predict overhead line thermal rating. In particular, such as [18,22], a distributed DTR architecture based on smart sensors network has been deployed to measure the conductor temperature and estimate the main variables governing the heat exchange between the conductor and its surroundings. The collected data is then transmitted to a central server for post-processing via a short-range communication system.

Although promising, the large-scale deployment of these facilities in modern transmission systems is still in its infancy, and several challenges remain to be addressed.

In particular, it is essential for the DTR-based sensors network to have a precise and common timing reference to determine the exact time when the samples of the conductor temperature and load current are taken. The synchronization of these samples to a common timing reference is achieved either internally or externally to the sensors network. The availability, reliability, and accuracy of the timing signals should be suitable for the thermal monitoring requirements [23]. The GPS-based timing signals have been explored to address these needs and offer certain benefits regarding wide area coverage and easy access to remote sites. However, the use of this technology requires the deployment of complex and costly hardware architectures which can affect the sensor costs.

Processing of sensors data is also a challenging task as the network should first identify the sensor that senses the maximum conductor temperature to compute the hotspot location. The corresponding sensor should then calculate the line ampacity and disseminate this information across the network. This computing process involves hierarchical architectures based on several network layers and a variety of data acquisition, transmission, concentration, and processing technologies. The centralized/hierarchical processing architecture for this problem requires the deployment of a central fusion center that acquires and processes all the sensors' measurements for all the monitored overhead lines. The large amount of data generated by the sensors, along with the field data produced by other grid sensors, should be transmitted to the central processing facilities by a wide-area communication network. However, due to the increasing number of field sensors in modern power systems, the amount of raw data is expected to rise quickly. Some research studies estimate an increase in the current data acquisition by about four orders of magnitude, leading to the saturation of existing centralized processing architectures. Consequently, TSOs must deal with communication bottlenecks, complex data management systems, and vulnerable centralized infrastructures.

Hence, researchers and developers of high-performance processing systems are re-examining various design elements and assumptions related to scale, reliability, heterogeneity, manageability, and system evolution over time as they strive to address issues in this area. These studies advocate a move toward a more distributed processing structure for power systems monitoring.

The selection of a more effective processing architecture for loadability assessment should be based on striking a balance between a cost-effective solution that provides better management and the one that offers higher performance and reliability. To this end, the potential use of ubiquitous computing and decentralized consensus protocols for addressing the thermal rating prediction problem have been proposed in several papers [24,25]. The primary concept is to enable distributed temperature sensors to implement the main functions required for DTR by cooperating in a coordinated manner, without requiring either explicit point-to-point message passing or routing protocols. This approach allows sensor networks to share local information, detect and respond to sensor failures, compute general functions of all the local variables sensed by the sensors, and transmit to a central processing facility only the most pertinent information about the actual and predicted loadability margins.

7.2 Self-organizing sensor network for DTR

In this section, we describe a DTR prediction framework centered on decentralized and self-organizing sensor networks [24,25]. The pivotal concept of this framework involves distributing the functions of sensing, synchronization, and processing of the data needed for DTR assessment across a self-organizing network of smart sensors, which are equipped with the distributed consensus protocols described in Chapter 2. These protocols enable the sharing of information among the smart sensors through iterative updates of each built-in oscillator state, based on weighted averages of its neighboring oscillators. This unique capability empowers smart sensors to synchronize their local measurements, and dynamically compute the conductor temperature in a fully decentralized manner, avoiding the requirement for a central fusion center aimed at collecting and processing all the sensor data. For this purpose, each smart sensor implements the following interactive tasks:

1. Time synchronization: achieving time synchronization among smart sensors is orchestrated by a decentralized protocol that adapts sensor clocks through mutual interactions within dynamic systems. This mechanism ensures local

clocks align to a common phase, despite differing individual oscillator frequencies, without necessitating a global or hierarchical synchronization infrastructure.

2. Sensing: time-synchronized smart sensors measure a set of environmental parameters (like environmental temperature, wind speed and direction, solar irradiance) by means of conventional sensors.
3. Conductor temperature computation: drawing from the measured weather variables and the conductor thermal parameters, the smart sensors infer the corresponding conductor temperature using an integrated thermal model.
4. Anomaly detection: to identify sensor failures or data outliers, smart sensors compute confidence intervals for the estimated conductor temperatures using a decentralized estimation protocol. This process empowers each smart sensor to determine its reliability as a "trustworthy node."
5. Critical span location: smart sensors designated as "trustworthy nodes" assess the highest conductor temperature along the entire line route (referred to as the hotspot temperature) and disseminate this information to all smart sensors. This is achieved through a decentralized and cooperative framework devised to address the max-consensus problem. This feature empowers the TSO to infer hotspot temperature insights by querying any smart sensor.

The conductor thermal parameters employed for conductor temperature computation are identified by a group of master nodes equipped with conductor temperature sensors. These nodes infer the values of key parameters governing conductor–environment heat exchange (e.g., wind speed) using an indirect parameter identification method, derived from measured conductor temperature, line current, and weather variables. Furthermore, these nodes assess the precision of the latest parameter set by solving an inherent thermal model. If the variance between predicted and measured conductor temperatures remains below a predetermined threshold, the current parameter set is confirmed as "validated"; otherwise, the node updates the calibrated parameter set.

The underlying theoretical principles and the operational technologies that enable these interactive functions will be expounded upon in subsequent sections.

7.2.1 Time-synchronization acquisition

The fundamental approach for designing an effective, robust, and economically viable framework for DTR assessment revolves around the synchronized acquisition of a set of environmental variables that define the conductor thermal behavior. This objective is achieved through the integration of local transducers within the smart sensors, ensuring that their measurements are seamlessly synchronized to a unified time reference. This complex task is accomplished through the implementation of two functions, namely the acquisition and the time-synchronization services. The former encompasses all essential tasks related to variables sensing, while the latter is focused on synchronizing all the smart sensors clocks:

$$\tau_i(t) = \alpha_i t + \beta_i \, \forall i \in [1, N] \tag{7.1}$$

to an unique "virtual" clock:

$$\tau_v(t) = \alpha_v\, t + \beta_v \tag{7.2}$$

where α_i and β_i represent the skew and offset coefficients of the ith local clock, respectively, while α_v, and β_v are the corresponding parameters of the virtual clock.

A decentralized approach founded on the average time synchronization protocol can be deployed to address this issue [26]. Incorporating this protocol empowers smart sensors to estimate the local clock parameters through the solution of the subsequent linear regression problem:

$$\hat{\tau}_i(t) = \hat{\alpha}_i\, t + \hat{\beta}_i \tag{7.3}$$

Here, $\hat{\alpha}_i$ and $\hat{\beta}_i$ represent the approximations of skew and offset coefficients for the ith smart sensor, respectively. This method of indirect estimation is performed by every smart sensor independently, solely through localized exchange of partial data, hence, avoiding the necessity of a central fusion center to gather and process all sensor data [27]. In particular, the main idea is to enable each smart sensor to periodically share its local time $\hat{\tau}_i(t)$ and the most recent estimations of its clock parameters (i.e., $\hat{\alpha}_i(k)$ and $\hat{\beta}_i(k)$) with its neighboring sensors N_i. Leveraging this information, each smart sensor refines its parameter estimations utilizing the subsequent gossip-based equations [28]:

$$\begin{cases} \hat{\alpha}_{ij}(k+1) = \rho\, \hat{\alpha}_{ij}(k) + (1-\rho)\frac{\tau_j(t_{k+1})-\tau_j(t_k)}{\tau_i(t_{k+1})-\tau_i(t_k)}\ \forall j \in N_i \\ \hat{\alpha}_i(k+1) = (1-\rho)\hat{\alpha}_i(k) + \rho\hat{\alpha}_{ij}(k+1)\hat{\alpha}_j(k)\ \forall j \in N_i \\ \hat{\beta}_i(k+1) = \hat{\beta}_i(k) + \rho\left(\hat{\alpha}_j(k)\tau_j(t_k) + \hat{\beta}_j(k) - \hat{\alpha}_i(k)\tau_i(t_k) + \hat{\beta}_i(k)\right)\ \forall j \in N_i \end{cases} \tag{7.4}$$

where $\rho \in [0,1]$ is a design parameter and, $hat\alpha_{ij}(0) = 1$, $hat\alpha_i(0) = 1$, and $hat\beta_i(0) = 1$. Assuming that certain assumptions regarding the communication network topology are met, encompassing a condition on the minimum connectivity degree, the approach built on gossiping quickly converge, hence, allowing the smart sensors to progressively align their clocks with an unique reference time, denoted as

$$\lim_{k\to\infty} \hat{\tau}_i(t_k) = \tau_v\ \forall i \in [1,N] \tag{7.5}$$

More specifically, since

$$\hat{\tau}_i(t) = \hat{\alpha}_i\alpha_i t + \hat{\alpha}_i\beta_i + \hat{\beta}_i \tag{7.6}$$

It follows that

$$\begin{aligned} &\lim_{k\to\infty} \hat{\alpha}_i(k)\alpha_i = \alpha_v \\ &\lim_{k\to\infty} (\hat{\alpha}_i(k)\beta_i + \hat{\beta}_i(k)) = \beta_v \end{aligned} \tag{7.7}$$

This allows the smart sensors to be synchronized to the unique clock $\tau_v(t)$, without requiring a global and centralized synchronization infrastructure.

7.2.2 Bad data detection

The accuracy of the periodically acquired time-synchronized measurements from smart sensors is routinely assessed to identify potential irregularities that might compromise the reliability of DTR computing. Fulfilling this fundamental function, the bad data detection service serves as a paramount tool for identifying and responding to sensor malfunctions or stochastic measurement inaccuracies. This tool is realized through the computation of key variables that characterize the spatial distribution of the conductor temperatures computed by the smart sensors and by the employment of conventional techniques for outlier detection. The orchestration of these calculations is executed via an information-spreading protocol, designed to establish consensus among smart sensors on the variables captured across the line route.

For this purpose, the built-in oscillators are initialized by the estimated conductor temperature T_i, which allows the sensors network to reach a consensus on the following equilibrium point:

$$\mu_T^* = \frac{\sum_{i=1}^N T_i}{N} \tag{7.8}$$

Which represents the spatial average of the conductor temperature measured by all the smart sensors. Moreover, initializing the built-in oscillators by the square of the estimated conductor temperature T_i^2 allows the sensors network to reach a consensus on the following equilibrium point:

$$\omega_c^* = \frac{\sum_{i=1}^N T_i^2}{N} \tag{7.9}$$

This global information allows the smart sensors to infer the spatial variance of the conductor temperature, as follows:

$$\sigma_T \leq T_i \leq^2 = \omega_c^* - \mu_T^* \tag{7.10}$$

The knowledge of μ_T^* and ω_c^* allows smart sensors to apply an anomaly detection method, such as assessing whether the measured conductor temperature falls within an appropriate tolerance interval:

$$\mu_T^* - 3\sigma_T \leq T_i \leq \mu_T^* + 3\sigma_T \tag{7.11}$$

This procedure empowers each individual smart sensor to gauge the congruity of the estimated conductor temperature, thereby enabling the identification of the "trustworthy nodes."

7.2.3 Conductor temperature calculation

Utilizing the measured meteorological data and the adjusted thermal parameters derived from the master nodes, this function estimates the conductor temperature by solving a dynamic thermal model.

The IEEE Standard 738-2012, which is a revised version of the IEEE Standard 738-2006, stands as the universally recognized protocol for evaluating the current–temperature relationship of bare overhead conductors. The temperature of the line conductor is influenced by many factors, which include conductor material, diameter, surface attributes, load current, and the local weather conditions. The impacts of all these factors are considered in this standard, which define detailed modeling technique aimed at computing the conductor's temperature based on the line current, or the capacity based on the utmost attainable operational temperature.

The thermal model is grounded in a fundamental first-order differential equation:

$$\frac{dT_{avg}}{dt} = \frac{1}{mC_p}[q_s + I^2R(T_{avg}) - q_c - q_r] \tag{7.12}$$

where

- T_{avg} is the conductor temperature.
- mC_p is the conductor heat capacity, which is defined as the product of its specific heat capacity, C_p, and the mass per unit of length, m.
- q_s is the solar heat gain rate per unit length.
- I is the line current.
- R is the electrical resistance of the conductor.
- q_c is the convected heat loss rate per unit length.
- q_r is the radiated heat loss rate per unit length.

In particular, q_c can be computed as function of the wind speed by using the following equations, which are valid for low, high, and zero wind speed, respectively:

$$q_c = \begin{cases} q_{c1} = K_{angle}(1.01 + 1.35\,Re^{0.52})k_f\,(T_s - T_a) \text{ for low wind speed} \\ q_{c2} = K_{angle}(0.754\,Re^{0.6})k_f\,(T_s - T_a) \text{ for high wind speed} \\ q_{cn} = 3.645\rho_f^{0.5}D_0^{0.75}(T_s - T_a)^{1.25} \text{ for zero wind speed} \end{cases} \tag{7.13}$$

where the variable K_{angle} can be computed as

$$K_{angle} = 1.194 - cos(\phi) + 0.194cos(2\phi) + 0.368sin(2\phi) \tag{7.14}$$

The angle ϕ depends by the wind direction, while the density, viscosity, and thermal conductivity of the air are estimated in function of the temperature T_{film} by using the following regression equations:

$$\begin{aligned} \mu_f &= \frac{1.458\;10^{-6}(T_{film}+273)^{1.5}}{T_{film}+383.4} \\ \rho_f &= \frac{1.293 - 1.525\;10^{-4}\;H_e + 6.379\;10^{-9}\;H_e^2}{1+0.00367\;T_{film}} \\ k_f &= 2.424\;10^{-2} + 7.477\;10^{-5}\;T_{film} - 4.407\;10^{-9}\;T_{film}^2 \end{aligned} \tag{7.15}$$

The terms q_r and q_s can be computed by using the following equations:

$$q_r = 17.8\,D_0\,\varepsilon\left[\left(\frac{T_s+273}{100}\right)^4 - \left(\frac{T_a+273}{100}\right)^4\right] \tag{7.16}$$

$$q_s = \alpha Q_{se} sin(\theta)A' \tag{7.17}$$

where θ is the angle of sun incidence, which can be computed in function of the solar altitude, H_c as

$$\theta = arcos(cos(H_c)cos(Z_c - Z_l)) \tag{7.18}$$

$$H_c = arcsin(cos(Lat)cos(\delta)cos(\omega) + sin(Lat)sin(\delta)) \tag{7.19}$$

$$\delta = 23.46\, sin\left(\frac{284+N}{365} + 360\right) \tag{7.20}$$

$$Z_c = C + arctan(\chi) \tag{7.21}$$

$$\chi = \frac{sin(\omega)}{sin(Lat)cos(\omega) - cos(Lat)tan(\delta)} \tag{7.22}$$

where the constant C is a function of hour angle and solar azimuth variable.

The corrected rate of solar heat gain in function of altitude, Q_{se}, can then be computed as

$$Q_{se} = K_{solar} Q_s \tag{7.23}$$

where

$$K_{solar} = A + B\, H_e + CH_e^2 \tag{7.24}$$

in which

$$\begin{cases} A = 1 \\ B = 1.148\ 10^{-4} \\ C = -1.108\ 10^{-8} \end{cases} \tag{7.25}$$

Q_s can be computed, for clear or industrial atmosphere, by the following regression equation in function of H_c:

$$Q_s = A + B\, H_c + CH_c^2 + DH_c^3 + EH_c^4 + FH_c^5 + GH_c^6 \tag{7.26}$$

where the constants A, B, C, D, E, F, G, and H are reported in [0].

Finally, the impact of thermal energy dissipation due to the Joule effect depends on the electrical resistance of the conductor at the average temperature T_{avg} and the square of the line current.

Conductor resistance exhibits variations based on factors such as system frequency, line current, and conductor temperature. In compliance with the IEEE 738-2012 Standard, alterations in electrical resistance are solely attributed to temperature fluctuations.

Upon providing an electrical resistance value for high temperatures, T_{high}, and another for low temperatures, T_{low}, it becomes viable to establish a linear model representing the relationship between electrical resistance and conductor temperature:

$$R(T_{avg}) = \left[\frac{R(T_{high} - R(T_{low}))}{T_{high} - T_{low}}\right](T_{avg} - T_{low}) + R(T_{low}) \tag{7.27}$$

Actually, the conductor resistance experiences a slightly swifter increase in comparison to what the linear model predicts in response to temperature changes. This signifies that if the resistance is computed for a temperature value T_{avg} lying between T_{low} and T_{high}, the resistance calculated using the linear model will be lower than the actual value. This conservative estimation is conducive to the conductor temperature computation. However, when the resistance is computed for T_{avg} exceeding T_{high}, the resultant resistance value will surpass the estimated value, leading to a non-conservative conductor temperature estimation. Moreover, it is important to account for variations in the resistance value due to the skin effect, magnetic core effect, and radial temperature gradient within the conductor.

7.2.4 Critical span location

Identifying the critical span within the observed overhead line entails calculating the most unfavorable thermal exchange scenario, as assessed by the smart sensors throughout the line route. To accomplish this, the smart sensors designated as "trustworthy nodes" compute the existing hotspot temperature by solving a decentralized max-consensus problem. To achieve this objective, the same computational scheme utilized in the bad data detection function is adopted. The solution of this maximum consensus problem allows all the dynamic oscillators to self-synchronizing, within a maximum of $N-1$ iterations, to the state of the max-leader—namely, the oscillator with the highest initial state.

7.2.5 Parameters calibration

The task of parameter calibration is executed within the master nodes, with the objective of determining the precise values of the conductor thermal model parameters. This objective is realized through the solution of a non-linear programming problem designed to infer the vector Γ, which include the adapted parameters of the conductor thermal model. The optimization process seeks to minimize the integral of the model prediction error within a specific time interval. This error encompasses the difference between the acquired conductor temperature and the corresponding estimation computed by solving the conductor thermal model. Mathematically, the issue can be defined as follows:

$$\min_{\Gamma,\Sigma} \frac{1}{T} \int_{t-T}^{T} w(T_c^m(\zeta)(T_c^m(\zeta) - T_c^e(\zeta,\Gamma,\Sigma)d\zeta) \tag{7.28}$$

In this optimization problem, the conductor thermal model parameters are the state variables, with the objective function being the integral of a weighted estimation error between the measured temperature $T_c^m(t)$ and the corresponding estimate $T_c^e(t)$.

The time interval T defines the temporal horizon assumed for parameter calibration. To effectively capture the time-evolving nature of thermal phenomena, it is recommended that T be equal to or exceed the lower time constant of parameter variation dynamics. Within this context, the dynamics of wind speed are recognized as a key factor in determining this value. Thus, an appropriate selection criterion should be established based on the anemometric features of the monitored location. The function $w(T_c^m(\zeta))$ is intended to adapt the error considering the operational conditions, by emphasizing enhanced estimation precision in scenarios with elevated line loading conditions.

To ascertain the accuracy of temperature calculations, the master nodes conduct continual evaluations to ensure the coherence of the most recent parameter set with the recorded conductor temperature. When this assessment yields a positive result, the current parameter set is validated. Otherwise, the calibration of the thermal parameter set is initiated.

7.3 Experimental results

To evaluate the advantages arising from applying the self-organizing framework for DTR assessment, detailed experimental results obtained on real operation scenario are here discussed. The focal transmission asset under analysis comprises a thermally restricted overhead line located in Southern Italy [4,24,25]. This overhead line plays a strategic role within the Italian power system due to the substantial wind energy output in the region, amounting to approximately 150—160 GWh/year. As such, enhancing the loading capacity of this asset facilitates the integration of wind power generators, preventing the power curtailments required for mitigating the effects of grid contingencies.

The experimental investigations were aimed at analyzing the decentralized sensors network performance in executing all the DTR functions. To this end, four prototype smart sensors were deployed along the overhead line route, spaced approximately 5 km apart. The sensors integrated along the line route feature a Rabbit BL2100 microcontroller coupled with a medium-range radio communication modem operating at 2.4 GHz, utilizing the IEEE 802.15.4 protocol. This facilitates information exchange between the microcontroller and its neighbors while also enabling the acquisition of conductor temperature through interaction with a Power Donut , which is a sophisticated sensor directly measuring the conductor temperature. All software routines embodying the local services were incorporated into the smart sensor via the Dynamic C development suite.

The trajectories of the embedded oscillators during the initiation of the synchronization services have been reported in Figure 7.1. These profiles confirmed the efficacy of the decentralized protocol in aligning conductor temperature measurements within approximately 30 s, thereby avoiding the need for external synchronization signals.

Following the evaluation of the decentralized time synchronization mechanism, we proceed to test the efficacy of the smart sensor network in addressing the bad data detection problem. For this purpose, the smart sensors first estimate the conductor temperature at the monitored span, obtaining the profiles in Figure 7.2, and then compute the mean and variance of the conductor temperature profile. This analysis was undertaken by solving two average consensus problems. The outcomes obtained for a 12-h time scenario have been summarized in Figures 7.3 and 7.4, illustrating the state trajectories of the embedded oscillators during the computation of the mean and variance of the conductor temperature for the analyzed time scenario.

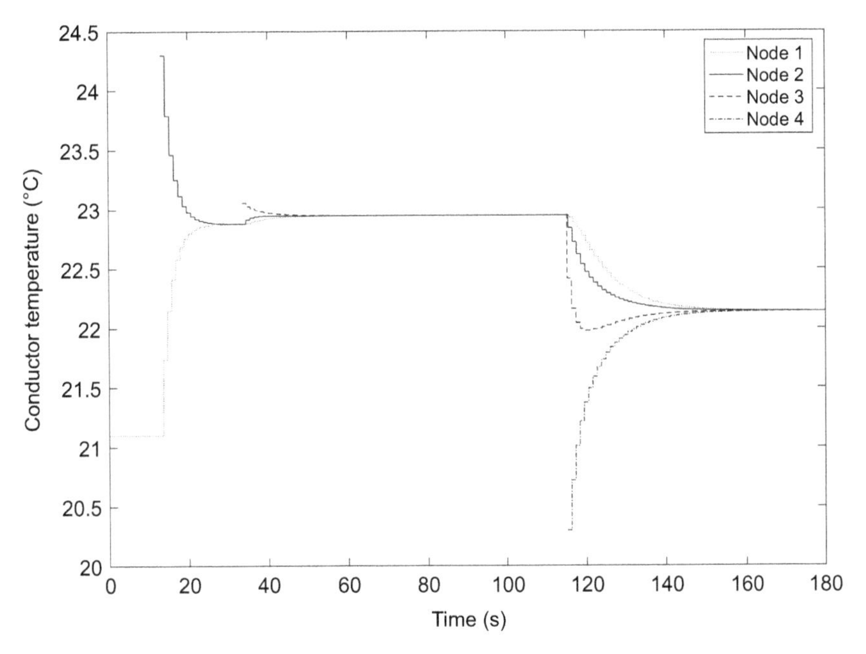

Figure 7.1 Time synchronization of the decentralized sensors network

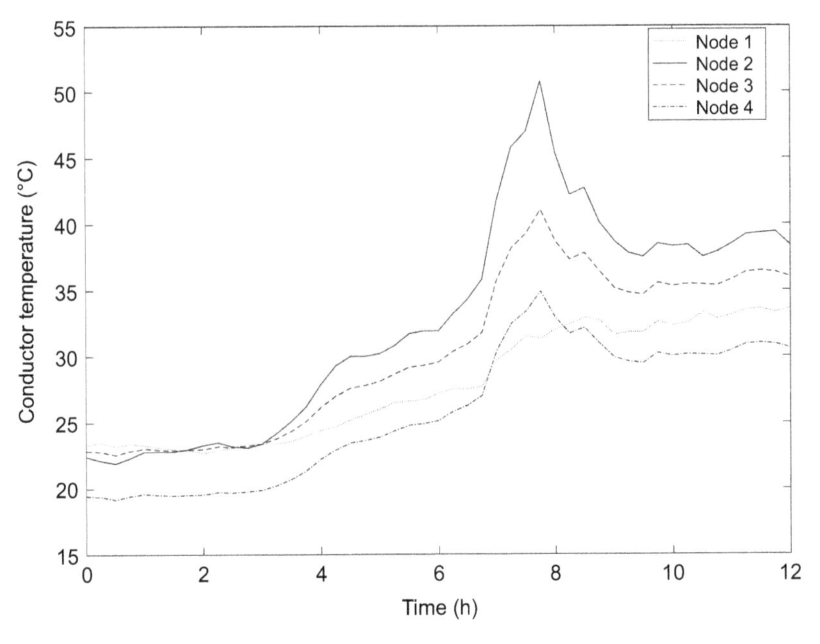

Figure 7.2 Conductor temperature profiles estimated by the smart sensors

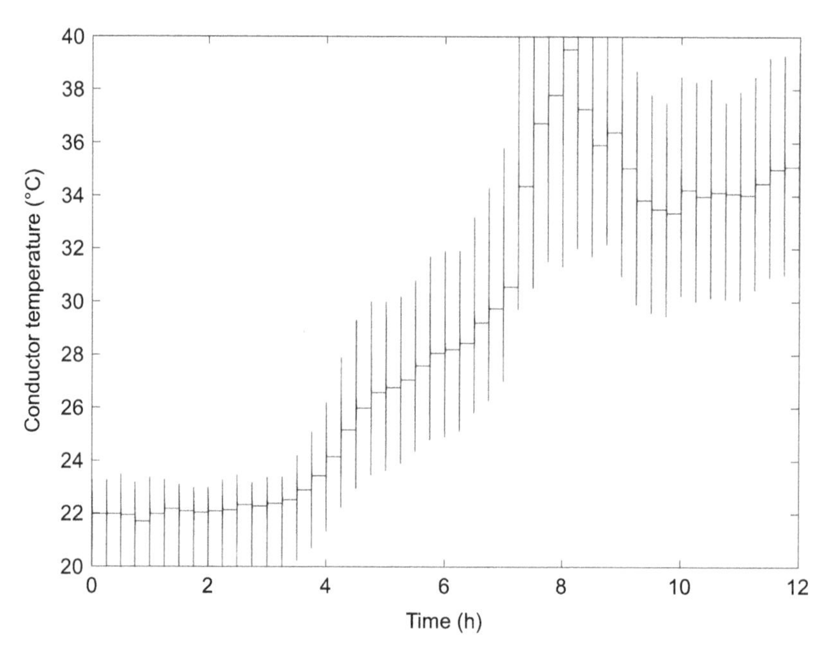

Figure 7.3 Trajectories of the state variables in the task of computing the average conductor temperature

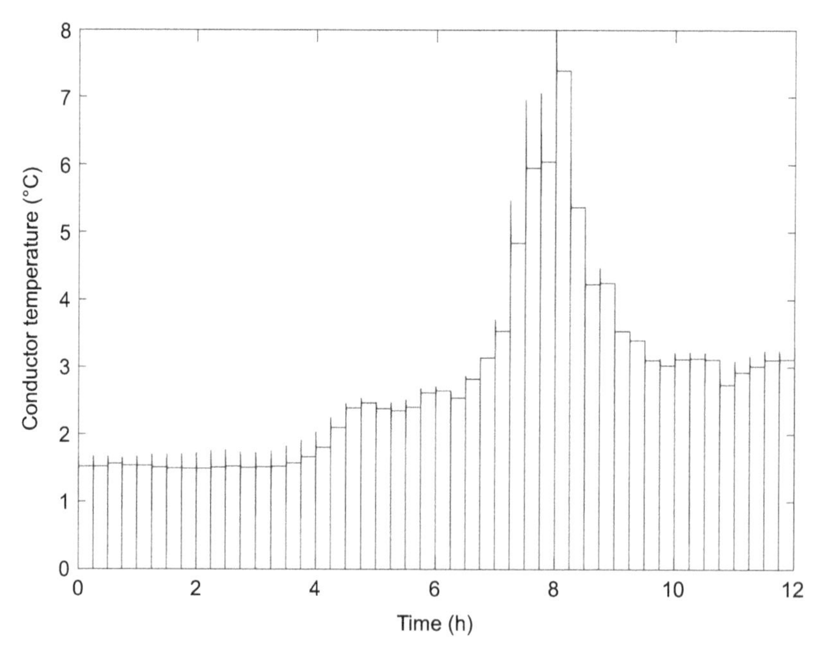

Figure 7.4 Trajectories of the state variables in the task of computing the spatial variance of the conductor temperature

Knowledge of these variables empowers the smart sensors to deduce tolerance intervals for the conductor temperature profile and to employ suitable outlier detection methods for identifying potential anomalies within the recorded data.

Building on these results, additional experimental activities focused on assessing the capabilities of the smart sensors in addressing the parameters calibration, temperature calculation, and loadability prediction challenges. Specifically, the parameters calibration service was executed by solving the problem presented in 7.28 in order to compute the conductor thermal model parameters (i.e. absorptivity and emissivity coefficients), the wind speed and direction, by using the measured conductor temperature, line current, and environmental temperature.

The critical role fulfilled by this service can be observed by examining Figure 7.5, depicting the temporal evolution of conductor temperature computed by the smart sensor without invoking the parameter calibration service (i.e., assuming nominal model parameters and measured wind speed and direction). A comparison with the actual profile measured by the Power Donut at a fixed line section reveals noticeable mismatches in various operational scenarios. This observation is further confirmed in Figure 7.6, which illustrates the histogram of prediction errors, indicating differences between computed and measured temperatures.

Analyzing these results underscores that the non-calibrated thermal model is more accurate when the load current is modest but becomes less precise (yielding a generally more conservative conductor temperature estimation) under higher loading conditions. This accentuates the importance of the parameters calibration service, which continually adapts the conductor thermal model parameters to precisely model the inherent time-varying nature of these intricate phenomena.

The advantages stemming from the periodic invocation of this service (e.g., every 10 min in this instance) are evident when examining Figure 7.7. This graph

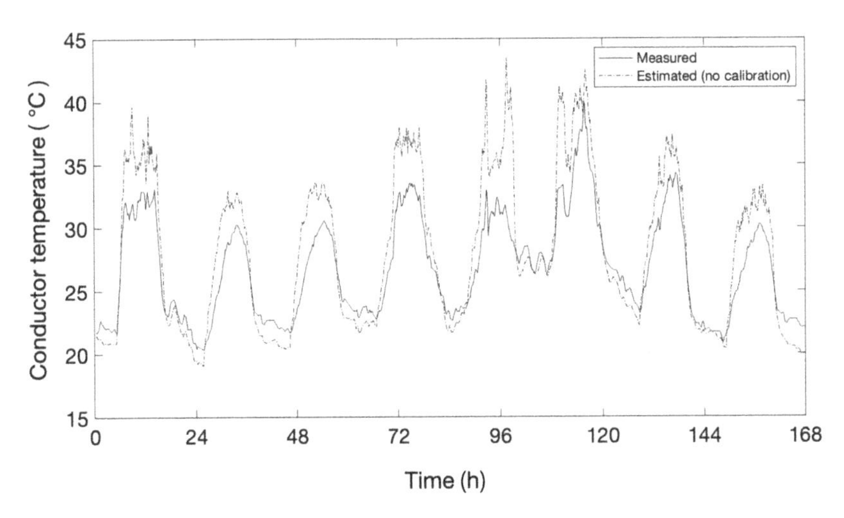

Figure 7.5 Conductor temperature estimated by node 3 using a not calibrated thermal parameters set

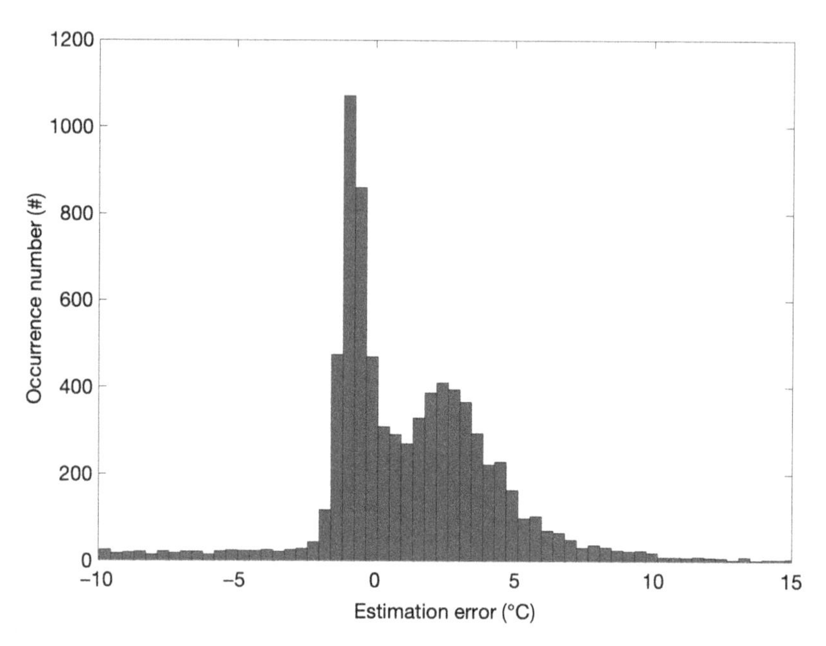

Figure 7.6 Histogram of the temperature estimation error

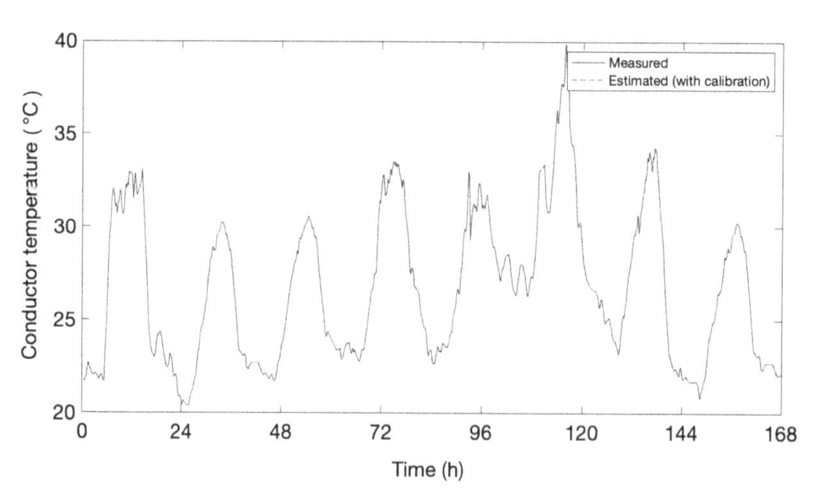

Figure 7.7 Conductor temperature estimated by node 3 using a calibrated thermal parameters set

presents the conductor temperature profile computed by the calibrated thermal model alongside the measured profile. In this case, the prediction error made by each smart sensor is roughly 10^{-2}°C.

With calibrated thermal model parameters at hand, the smart sensors can forecast conductor temperature across various time-frames by invoking the temperature calculation service, equipped with the calibrated parameter set, actual line current, and assuming the persistence of environmental variables. Evaluating the error between conductor temperatures predicted by the thermal model and the measured values enables an assessment of temperature prediction accuracy. This error pertains to six prediction steps (1—6) spanning from 10 min to 1 h. At step 0, the thermal model is calibrated, and recalibration follows at the end of step 6 for the next prediction windows. The results for a typical daily operation, observed through the node 3, are summarized in Figure 7.8.

As expected, the forecasting error increases as the prediction horizon extends, as the performance of the persistence model used to predict the evolution of environmental variables (particularly wind speed) tends to degrade over longer horizons. Based on these experimental outcomes, we can conclude that the temperature calculation service accuracy is suitable for very-short-term loadability analysis, namely for forecasting horizons no longer than 1–2 h.

Building on these findings, the smart sensor located on the critical span calculates the load capacity curve by iteratively invoking the temperature calculation service for a range of hypothetical load currents. This process considers the calibrated thermal model parameters, actual conductor thermal state, and forecasted environmental variables. To underscore the loadability curve sensitivity to operational conditions, results for two distinct scenarios—a lightly loaded and fully loaded condition—are shown in Figures 7.9 and 7.10, respectively. These figures benchmark the loadability curves computed by the smart sensor with those derived from

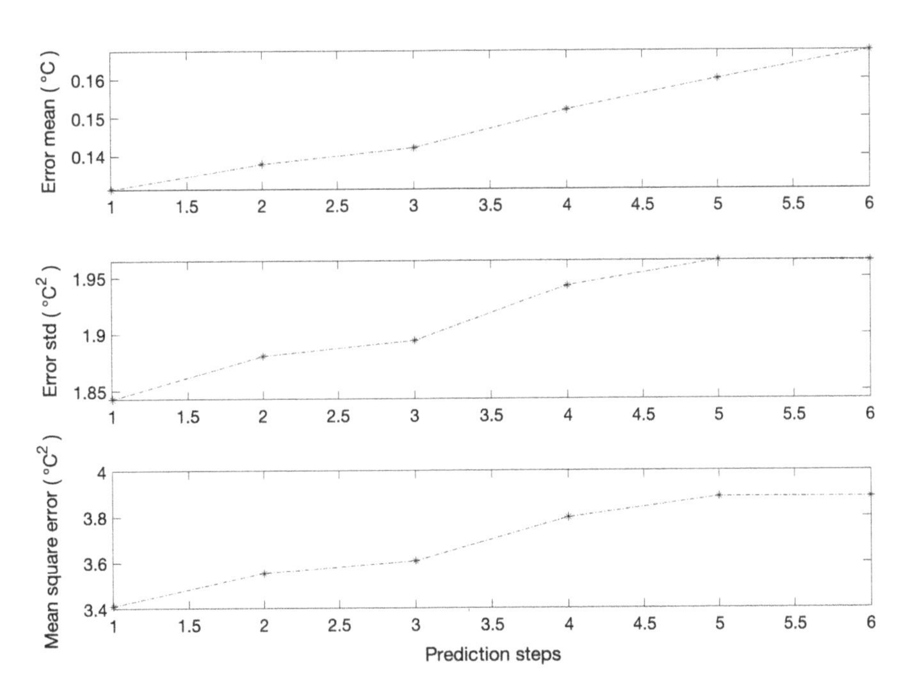

Figure 7.8 Evolution of the forecasting errors for multiple time horizons

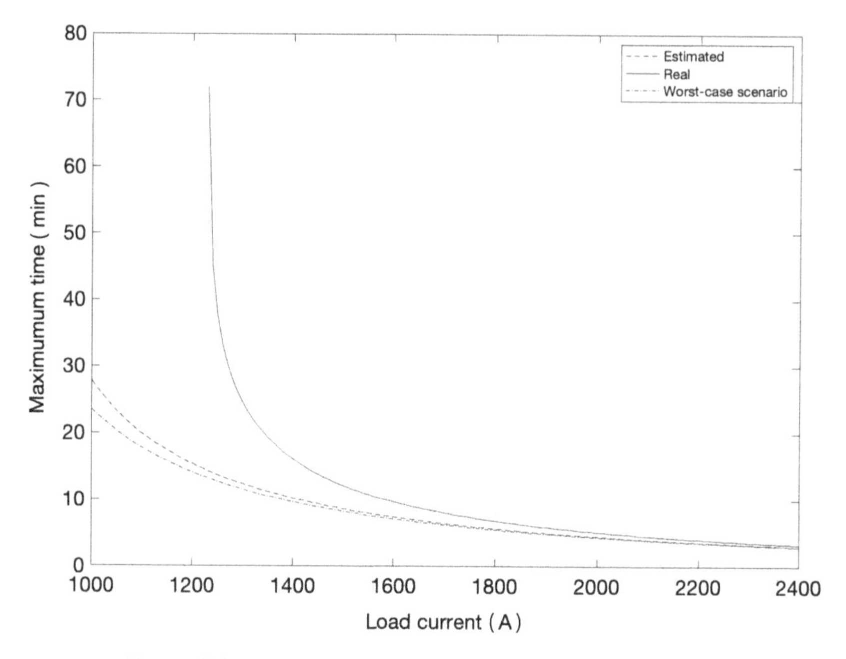

Figure 7.9 Load capability curves for highly loaded line

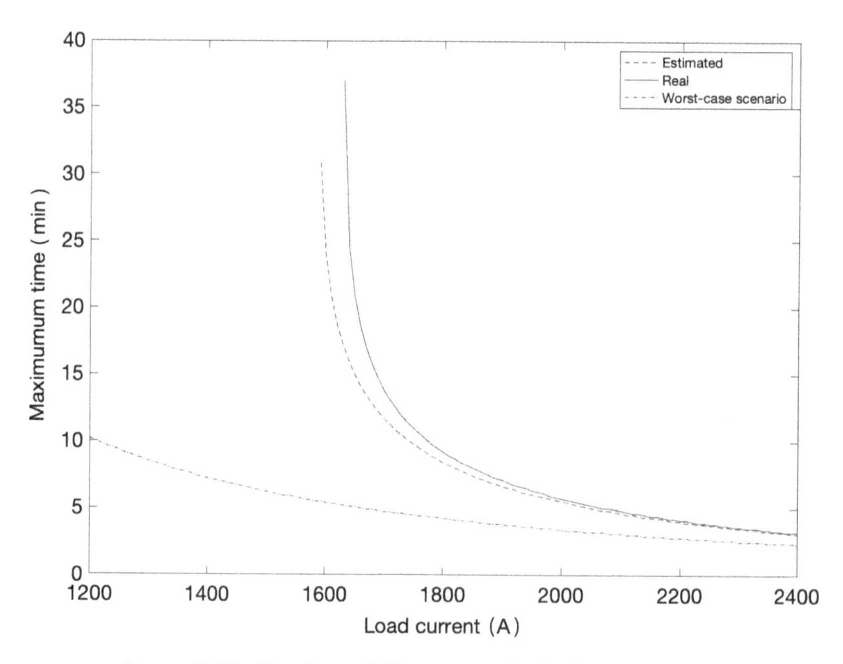

Figure 7.10 Load capability curves for lightly loaded line

actual measured input parameters (i.e., measured wind velocity and direction) and the worst-case estimation (zero wind speed, perpendicular to conductor axis).

Analyzing these results reveals that load capacity curves computed by the smart sensor are conservative in relation to the actual curve, yielding a more precise estimation of the real loading margins compared to the worst-case scenario. This is particularly evident in Figure 7.10, where the curve computed by the smart sensor closely aligns with the real curve while diverging from the worst-case curve. These experimental results confirm the time-varying nature of the DTR prediction problem. They also emphasize the crucial role of adaptive computing in achieving dependable and accurate insight into transmission asset true capability margins. Importantly, all these results were achieved via a fully decentralized framework, with smart sensors gathering and processing all local measurements, bypassing the necessity for a central data fusion center. This decentralized framework lends the DTR architecture scalability, self-organization, and distribution.

References

[1] Deb AK. *Powerline Ampacity System: Theory, Modeling and Applications*. 21st ed. Press C, editor. Boca Raton: CRC Press; 2000.

[2] Piccolo A, Vaccaro A, and Villacci D. Thermal rating assessment of overhead lines by affine arithmetic. *Electric Power Systems Research*. 2004;71(3):275–283.

[3] Alvarez DL, Faria da Silva F, Mombello EE, *et al*. An approach to dynamic line rating state estimation at thermal steady state using direct and indirect measurements. *Electric Power Systems Research*. 2018;163:599–611. Advances in HV Transmission Systems.

[4] Carlini EM, Giannuzzi GM, Pisani C, *et al*. Experimental deployment of a self-organizing sensors network for dynamic thermal rating assessment of overhead lines. *Electric Power Systems Research*. 2018;157:59–69.

[5] Mahin AU, Islam SN, Ahmed F, *et al*. Measurement and monitoring of overhead transmission line sag in smart grid: a review. *IET Generation, Transmission & Distribution*. 2022;16(1):1–18. Available from: https://ietresearch.onlinelibrary.wiley.com/doi/abs/10.1049/gtd2.12271.

[6] Pepiciello A, Coletta G, Vaccaro A, *et al*. The role of learning techniques in synchrophasor-based dynamic thermal rating. *International Journal of Electrical Power & Energy Systems*. 2020;115:105435. Available from: https://www.sciencedirect.com/science/article/pii/S0142061519307781.

[7] Tumelo-Chakonta C and Kopsidas K. A probabilistic indicator of the optimal operator action time under short-time emergency line loadings. In: *2015 IEEE Eindhoven PowerTech*; 2015. p. 1–6.

[8] Bucher MA, Vrakopoulou M, and Andersson G. Probabilistic *N*–1 security assessment incorporating dynamic line ratings. In: *2013 IEEE Power & Energy Society General Meeting*; 2013. p. 1–5.

[9] IEEE Standard for Calculating the Current-Temperature Relationship of Bare Overhead Conductors. IEEE Std 738-2012 (Revision of IEEE Std 738-2006 - Incorporates IEEE Std 738-2012 Cor 1-2013). 2013; p. 1–72.

[10] Pepiciello A. Flexibility solutions for the integration of variable renewable energy sources in power systems. PhD Thesis, University of Sannio; 2021.

[11] Poli D, Lutzemberger G, and Pelacchi P. Electromechanical dynamics of overhead electric power line conductors analysed through the Modelica language models. In: *2016 Power Systems Computation Conference (PSCC)*; 2016. p. 1–7.

[12] Motlis Y, Barrett JS, Davidson GA, *et al.* Limitations of the ruling span method for overhead line conductors at high operating temperatures. *IEEE Transactions on Power Delivery*. 1999;14(2):549–560.

[13] Chu RF. On selecting transmission lines for dynamic thermal line rating system implementation. *IEEE Transactions on Power Systems*. 1992;7(2): 612–619.

[14] Matus M, Saez D, Favley M, *et al.* Identification of critical spans for monitoring systems in dynamic thermal rating. *IEEE Transactions on Power Delivery*. 2012;27(2):1002–1009.

[15] CIGRE. Thermal behaviour of overhead conductors. In: Technical Brochure 207; 2002.

[16] CIGRE. Guide for thermal rating calculations of overhead lines. In: Technical Brochure 601; 2014.

[17] Wan H, McCalley JD, and Vittal V. Increasing thermal rating by risk analysis. *IEEE Transactions on Power Systems*. 1999;14(3):815–828.

[18] Vaccaro A and Villacci D. A cooperative smart sensor network for dynamic loading of overhead Lines. In: *2011 CIGRE' International Symposium on The Electric Power System of the Future – Integrating Supergrids and Microgrids*; 2011.

[19] Coletta G, Vaccaro A, and Villacci D. A review of the enabling methodologies for PMUs-based dynamic thermal rating of power transmission lines. *Electric Power Systems Research*. 2017;152:257–270.

[20] Palma PD, Collin A, Caro FD, *et al.* The role of fiber optic sensors for enhancing power system situational awareness: a review. *Smart Grids and Sustainable Energy*. 2024;9:1–26.

[21] Yang Y, Harley RG, Divan D, *et al.* Real-time dynamic thermal rating evaluation of overhead power lines based on online adaptation of echo state networks. In: *International Conference on Energy Conversion Congress and Exposition*; 2010.

[22] Vaccaro A and Villacci D. An adaptive smart sensor network for overhead lines thermal rating prediction. *International Journal of Emerging Electric Power Systems*. 2008;9(4):1–18.

[23] Vaccaro A, Zobaa AF, and Formato G. Vulnerability analysis of satellite based synchronized smart grids monitoring systems. *Electric Power Components and Systems*. 2014;42(3–4):408–417.

[24] Villacci D, Orrù L, Gasparotto F, *et al.* Experimental assessment of cooperative sensors network-based dynamical thermal rating: the first evidences from the H2020 OSMOSE Project. In: *2021 AEIT International Annual Conference (AEIT)*; 2021. p. 1–5.
[25] Villacci D, Gasparotto F, Orrú L, *et al.* Congestion management in Italian HV grid using novel dynamic thermal rating methods: first results of the H2020 European project Osmose. In: *2020 AEIT International Annual Conference (AEIT)*; 2020. p. 1–6.
[26] Ghabcheloo R, Aguiar AP, Pascoal A, *et al.* Synchronization in multi-agent systems with switching topologies and non-homogeneous communication delays. In: *2007 46th IEEE Conference on Decision and Control*; 2007. p. 2327–2332.
[27] Maggs MK, O'Keefe SG, and Thiel DV. Consensus clock synchronization for wireless sensor networks. *IEEE Sensors Journal*. 2012;12(6):2269–2277.
[28] Schenato L and Fiorentin F. Average TimeSynch: a consensus-based protocol for clock synchronization in wireless sensor networks. *Automatica*. 2011;47(9):1878–1886.

Index